NATURAL GAS TECHNOLOGY AND BUSINESS

NATURAL GAS TECHNOLOGY AND BUSINESS

Abhijeet Mishra

RANDOM
RANDOM PUBLICATIONS
NEW DELHI - 110 002 (INDIA)

Natural Gas Technology and Business

ISBN 978-93-51116-66-0

Published in 2015 in India by

RANDOM PUBLICATIONS

4376-A/4B, Gali Murari Lal, Ansari Road
New Delhi-110 002
Phone: +9111-43580356, 23289044
E-mail: randomexports@gmail.com; sales@randompublications.com; info@randompublications.com

Reprinted 2026

Type Setting by: Friends Media, Delhi-110089

Preface

Natural gas is an extremely valued resource in our society. It is not surprising, then, that the natural gas industry generates a great deal of commerce worldwide. In the 19th century, natural gas was usually obtained as a by-product of producing oil, since the small, light gas carbon chains came out of solution as the extracted fluids underwent pressure reduction from the reservoir to the surface, similar to uncapping a bottle of soda where the carbon dioxide effervesces. Unwanted natural gas was a disposal problem in the active oil fields. If there was not a market for natural gas near the wellhead it was virtually valueless since it had to be piped to the end user. Technological innovation in the exploration and production sector has equipped the industry with the equipment and practices necessary to continually increase the production of natural gas to meet rising demand. Because of its low density, it is not easy to store natural gas or to transport it by vehicle. Natural gas pipelines are impractical across oceans.

Gas is turned into liquid at a liquefaction plant, and is returned to gas form at regasification plant at the terminal. Shipborne regasification equipment is also used. LNG is the preferred form for long distance, high volume transportation of natural gas, whereas pipeline is preferred for transport for distances up to 4,000 km (2,485 mi) over land and approximately half that distance offshore. CNG is transported at high pressure, typically above 200 bars. Compressors and decompression equipment are less capital intensive and may be economical in smaller unit sizes than liquefaction/ regasification plants. Natural gas trucks and carriers may transport natural gas directly to end-users, or to distribution points such as pipelines. Floating Liquefied Natural Gas (FLNG) is an innovative technology designed to enable the development of offshore gas resources that would otherwise remain untapped because due to environmental or economic factors it is nonviable to develop them via a land-based LNG operation. In mines, where methane seeping from rock formations has no odour, sensors are used, and mining apparatus such as the Davy lamp has been specifically developed to avoid ignition sources. Some gas fields yield sour gas containing hydrogen sulfide (H_2S). This untreated gas is toxic. Amine gas treating, an industrial scale process which removes acidic gaseous components, is often used to remove hydrogen sulfide from natural gas. Extraction of natural gas (or oil) leads to decrease in pressure in the reservoir. Such decrease in pressure in turn may result in subsidence, sinking of the ground above. Subsidence may affect ecosystems, waterways, sewer and water supply systems, foundations, and so on.

The present reference book on Natural Gas Technology is a comprehensive work covering almost all the facets of technology and business issue regarding Natural Gas which is playing a stellar role in meeting the world's energy needs. This book will serve as a reference for all engineers and professionals in the energy business.

I thank all members of my team who have helped in the preparation of the book. My special thanks go to "Random Publications" who have published the book.

— Abhijeet Mishra

Contents

Chapter 1

Introduction

Natural Gas is one of the principle sources of energy for many of our day-to-day needs and activities. This section provides an overview of some of the characteristics of this vital gas. The links [to the left] lead you to a number of areas of interest surrounding natural gas as a fossil fuel.

Background

Natural gas is a vital component of the world's supply of energy. It is one of the cleanest, safest, and most useful of all energy sources. Despite its importance, however, there are many misconceptions about natural gas. For instance, the word 'gas' itself has a variety of different uses, and meanings. When we fuel our car, we put 'gas' in it. However, the gasoline that goes into your vehicle, while a fossil fuel itself, is very different from natural gas. The 'gas' in the common barbecue is actually propane, which, while closely associated and commonly found in natural gas, is not really natural gas itself. While commonly grouped in with other fossil fuels and sources of energy, there are many characteristics of natural gas that make it unique. Below is a bit of background information about natural gas, what it is exactly, how it is formed, and how it is found in nature.

What is Natural Gas?

Natural gas, in itself, might be considered an uninteresting gas - it is colorless, shapeless, and odorless in its pure form. Quite uninteresting - except that natural gas is combustible, abundant in the United States and when burned it gives off a great deal of energy and few emissions. Unlike other fossil fuels, natural gas is clean burning and emits lower levels of potentially harmful byproducts into the air. We require energy constantly, to heat our homes, cook our food, and generate our electricity. It is this need for energy that has elevated natural gas to such a level of importance in our society, and in our lives.

Natural gas is a combustible mixture of hydrocarbon gases. While natural gas is formed primarily of methane, it can also include ethane, propane, butane and pentane. The composition of natural gas can vary widely, but below is a chart outlining the typical makeup of natural gas before it is refined.

Typical Composition of Natural Gas

Methane	CH_4	70-90%
Ethane	C_2H_6	0-20%
Propane	C_3H_8	
Butane	C_4H_{10}	
Carbon Dioxide	CO_2	0-8%
Oxygen	O_2	0-0.2%
Nitrogen	N_2	0-5%
Hydrogen sulphide	H_2S	0-5%
Rare gases	A, He, Ne, Xe	trace

In its purest form, such as the natural gas that is delivered to your home, it is almost pure methane. Methane is a molecule made up of one carbon atom and four hydrogen atoms, and is referred to as CH4. The distinctive "rotten egg" smell that we often associate with natural gas is actually an odorant called mercaptan that is added to the gas before it is delivered to the end-user. Mercaptan aids in detecting any leaks.

Ethane, propane, and the other hydrocarbons commonly associated with natural gas have slightly different chemical formulas. Click on the link for a closer look into the combustion of methane. Natural gas is considered 'dry' when it is almost pure methane, having had most of the other commonly associated hydrocarbons removed. When other hydrocarbons are present, the natural gas is 'wet'.

Natural gas is considered 'dry' when it is almost pure methane, having had most of the other commonly associated hydrocarbons removed. When other hydrocarbons are present, the natural gas is 'wet.'

Natural gas has many uses, residentially, commercially, and industrially. Found in reservoirs underneath the earth, natural gas is often associated with oil deposits. Production companies search for evidence of these reservoirs by using sophisticated technology that helps to find the location of the natural gas, and drill wells in the earth where it is likely to be found. Click on the link to learn more about the new technologies and their environmental impact. Once brought from underground, the natural gas is refined to remove impurities such as water, other gases, sand, and other compounds. Some hydrocarbons are removed and sold separately, including propane and butane. Other impurities are also removed, such as hydrogen sulfide (the refining of which can produce sulfur, which is then also sold separately). After refining, the clean natural gas is transmitted through a network of pipelines, thousands of miles of which exist in the United States alone. From these pipelines, natural gas is delivered to its point of use.

Natural gas can be measured in a number of different ways. As a gas, it can be measured by the volume it takes up at normal temperatures and pressures, commonly expressed in cubic feet. Production and distribution companies commonly measure natural gas in thousands of cubic feet (Mcf), millions of cubic feet (MMcf), or trillions of cubic feet (Tcf). While measuring by volume is useful, natural gas can also be measured as a source of energy. Like other forms of energy, natural gas is commonly measured and expressed in British thermal units (Btu). One Btu is the amount of natural gas that will produce enough energy to heat one pound of water by one degree at normal pressure. To give an idea, one cubic foot of natural gas contains about 1,027 Btus. When natural gas is delivered to a residence, it is measured by the gas utility in 'therms' for billing purposes. A therm is equivalent to 100,000 Btu, or just over 97 cubic feet, of natural gas.

The Formation of Natural Gas

Natural gas is a fossil fuel. Like oil and coal, this means that it is, essentially, the remains of plants and animals and microorganisms that lived millions and millions of years ago. But how do these once living organisms become an inanimate mixture of gases?

There are many different theories as to the origins of fossil fuels. The most widely accepted theory says that fossil fuels are formed when organic matter (such as the remains of a plant or animal) is compressed under the earth, at very high pressure for a very long time. This is referred to as thermogenic methane. Similar to the formation of oil, thermogenic methane is formed from organic particles that are covered in mud and other sediment. Over time, more and more sediment and mud and other debris are piled on top of the organic matter. This sediment and debris puts a great deal of pressure on the organic matter, which compresses it.

This compression, combined with high temperatures found deep underneath the earth, breaks down the carbon bonds in the organic matter. As one gets deeper and deeper under the earth's crust, the temperature gets higher and higher. At low temperatures (shallower deposits), more oil is produced relative to natural gas. At higher temperatures, however, more natural gas is created, as opposed to oil. That is why natural gas is usually associated with oil in deposits that are 1 to 2 miles below the earth's crust. Deeper deposits, very far underground, usually contain primarily natural gas, and in many cases, pure methane.

Natural gas can also be formed through the transformation of organic matter by tiny microorganisms. This type of methane is referred to as biogenic methane. Methanogens, tiny methane-producing microorganisms, chemically break down organic matter to produce methane. These microorganisms are commonly found in areas near the surface of the earth that are void of oxygen.

These microorganisms also live in the intestines of most animals, including humans. Formation of methane in this manner usually takes place close to the surface of the earth, and the methane produced is usually lost into the atmosphere. In certain circumstances, however, this methane can be trapped underground, recoverable as natural gas. An example of biogenic methane is landfill gas. Waste-containing landfills produce a relatively large amount of natural gas from the decomposition of the waste materials that they contain. New technologies are allowing this gas to be harvested and used to add to the supply of natural gas.

A third way in which methane (and natural gas) may be formed is through abiogenic processes. Extremely deep under the earth's crust, there exist hydrogen-rich gases and carbon molecules. As these gases gradually rise towards the surface of the earth, they may interact with minerals that also exist underground, in the absence of oxygen. This interaction may result in a reaction, forming elements and compounds that are found in the atmosphere (including nitrogen, oxygen, carbon dioxide, argon, and water). If these gases are under very high pressure as they move toward the surface of the earth, they are likely to form methane deposits, similar to thermogenic methane.

Natural Gas Under the Earth

Although there are several ways that methane, and thus natural gas, may be formed, it is usually found underneath the surface of the earth. As natural gas has a low density, once formed it will rise toward the surface of the earth through loose, shale type rock and other material. Some of this methane will simply rise to the surface and dissipate into the air. However, a great deal of this methane will rise up into geological formations that 'trap' the gas under the ground. These formations are made up of layers of porous, sedimentary rock (kind of like a sponge that soaks up and contains the gas), with a denser, impermeable layer of rock on top.

This impermeable rock traps the natural gas under the ground. If these formations are large enough, they can trap a great deal of natural gas underground, in what is known as a reservoir. There are a number of different types of these formations, but the most common is created when the impermeable sedimentary rock forms a 'dome' shape, like an umbrella that catches all of the natural gas that is floating to the surface.

There are a number of ways that this sort of 'dome' may be formed. For instance, faults are a common location for oil and natural gas deposits to exist. A fault occurs when the normal sedimentary layers 'split' vertically, so that impermeable rock shifts down to trap natural gas in the more permeable limestone or sandstone layers. Essentially, the geological formation, which layers impermeable rock over more porous, oil and gas rich sediment, has the potential to form a reservoir.

The picture below shows how natural gas and oil can be trapped under impermeable sedimentary rock, in what is known as an anticlinal formation. To successfully bring these fossil fuels to the surface, a hole must be drilled through the impermeable rock to release the fossil fuels under pressure. Note that in reservoirs that contain oil and gas, the gas, being the least dense, is found closest to the surface, with the oil beneath it, typicall followed by a certain amount of water. With natural gas trapped under the earth in this fashion, it can be recovered by drilling a hole through the impermeable rock. Gas in these reservoirs is typically under pressure, allowing it to escape from the reservoir on its own. In addition to being found in a traditional reservoir such as the one shown above, natural gas may also be found in other 'unconventional' formations.

Uses of Natural Gas

For hundreds of years, natural gas has been known as a very useful substance. The Chinese discovered a very long time ago that the energy in natural gas could be harnessed, and used to heat water. In the early days of the natural gas industry, the gas was mainly used to light streetlamps, and the occasional house. However, with much improved distribution channels and technological advancements, natural gas is being used in ways never thought possible. There are so many different applications for this fossil fuel that it is hard to provide an exhaustive list of everything it is used for. And no doubt, new uses are being discovered all the time.

To learn more about technological advancements in the natural gas industry. Natural gas has many applications, commercially, in your home, in industry, and even in the transportation sector! While the uses described here are not exhaustive, they may help to show just how many things natural gas can do.

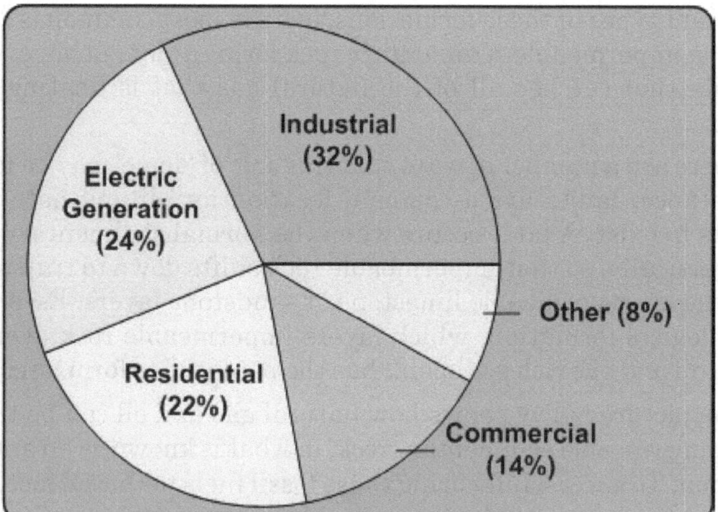

Figure: *Natural Gas Use By Sector*
Source: EIA - Annual Energy Outlook 2002

According to the Energy Information Administration, energy from natural gas accounts for 24 percent of total energy consumed in the United States, making it a vital component of the nation's energy supply. For more detailed information on the demand for and supply of energy, and natural gas, including forecasts and outlooks.

Natural gas is used across all sectors, in varying amounts. The graph below gives an idea of the proportion of natural gas use per sector. The industrial sector accounts for the greatest proportion of natural gas use in the United States, with the residential sector consuming the second greatest quantity of natural gas.

How Much Natural Gas is There?

There is an abundance of natural gas in North America, but it is a non-renewable resource, the formation of which takes thousands and possibly millions of years. Therefore, understanding the availability of our supply of natural gas is important as we increase our use of this fossil fuel.

This section will provide a framework for understanding just how much natural gas there is in the ground available for our use, as well as links to the most recent statistics concerning the available supply of natural gas.

As natural gas is essentially irreplaceable (at least with current technology), it is important to have an idea of how much natural gas is left in the ground for us to use. However, this becomes complicated by the fact that no one really knows exactly how much natural gas exists until it is extracted. Measuring natural gas in the ground is no easy job, and it involves a great deal of inference and estimation. With new technologies, these estimates are becoming more and more reliable; however, they are still subject to revision.

A common misconception about natural gas is that we are running out, and quickly. However, this couldn't be further from the truth. Many people believe that price spikes, such as were seen in the 1970's, and in the early to mid-2000s, indicate that we are running out of natural gas. The two aforementioned periods of high prices were not caused by waning natural gas resources - rather, there were other forces at work in the marketplace. In fact, there is a vast amount of natural gas estimated to still be in the ground. In order to better understand exactly what these estimates mean, and their importance, it is useful first to learn a bit of industry terminology for the different types of estimates.

U.S. Natural Gas Resource Estimates

Below are three estimates of natural gas reserves in the United States. It is important to note that analysts use different methodologies and systems of classification in various estimates. There is no single way that every industry player quantifies estimates of natural gas.

Therefore, it is important to delve into the assumptions and methodology behind each study to gain a complete understanding of the estimate itself. The first, compiled by the Energy Information Administration (referred to as the EIA), estimates that there are 2,543 trillion cubic feet (Tcf) of technically recoverable natural gas in the United States.

This includes undiscovered, unproved, and unconventional natural gas. To learn more about these classifications. As can be seen from the table, proved reserves make up a very small proportion of the total recoverable natural gas resources in the U.S. The most recent EIA data on proved reserves in the U.S. can be found here.

The following table includes an estimate of natural gas resources compiled by the National Petroleum Council in its 2007 report Facing the Hard Truths About Energy. In their study, the National Petroleum Council estimated there were 1,451 Tcf remaining in the U.S. According to this 2007 report, advances in technology could bring natural gas resources to 1,887 Tcf by 2017.

Since then, natural gas resources have dramatically changed because of the ability to extract natural gas from shale rock found in more than 20 states in the U.S. The National Petroleum Council has undertaken an updated analysis of resources. This *North American Resources Study* was scheduled to be released in March of 2011 but is not yet available.

Natural Gas Resources

(Trillion Cubic Feet)

Traditional Resources	
Lower 48 States	
On shore	764
Off shore	384
Alaska	303
Total	**1451**

Source: National Petroleum Council, Facing the Hard Truths about Energy, 2007.

In addition to the three estimates above, there are a myriad of different industry participants that formulate their own estimates regarding natural gas supplies. While this leads to a wealth of information, it also leads to a number of difficulties. Each estimate is based on a different set of assumptions, completed with different tools and even referred to with different language. It is thus difficult to get a definitive answer to the question of how much natural gas exists. In addition, since these are all essentially educated guesses as to the amount of natural gas in the earth, there are constant revisions being made. New technology, combined with increased knowledge of particular areas and reservoirs mean that these estimates are in a constant state of flux. Further complicating the scenario is the fact that there are no universally accepted definitions for the terms that are used differently by geologists, engineers, accountants, and others.

The graph, using data recorded by EIA shows the change in proved natural gas reserves in the United States from 1979 to 2009. As can be seen, the levels of proved reserves in the United States have changed significantly over the past 10 years. If the additions to proved reserves in a given year are larger than the subtractions from production, then proved reserves will increase, and vice versa. Usually, however, the additions are close enough to the subtractions to maintain a relatively constant level of proved reserves.

Where are These Reserves?

Most of the natural gas that is found in North America is concentrated in relatively distinct geographical areas, or basins. Given this distribution of natural gas deposits, those states which are located on top of a major basin have the highest level of natural gas reserves. As can be seen from the map below, U.S. natural gas reserves historically have been concentrated around Texas and the Gulf of Mexico. With the recent onset of shale production the number of states across the U.S with significant resources has increased, with major shale gas producing states including New York, Pennsylvania, Arkansas and Oklahoma, as well as Texas and Louisiana.

The map to the right gives a general impression of where most of the proved natural gas resources are in the United States.

World Natural Gas Reserves

The EIA, in conjunction with the *Oil and Gas Journal* and *World Oil* publications, estimates world proved natural gas reserves to be around 6,609 Tcf. As can be seen from the graph, most of these reserves are located in the Middle East with 2,658 Tcf, or 40 percent of the world total, and Europe and the Former U.S.S.R. with 2,331, or 35 percent of total world reserves. The United States, by this calculation, possesses slightly over 4 percent of the world total natural gas reserves.

Now that the methods of reserve estimation and the most recent data concerning North America and the world's supply of natural gas have been discussed, proceed to learn about the many uses of this vital gas.

Natural gas is nothing new: In fact, most of the natural gas that is brought out from under the ground is millions and millions of years old. However, it was not until recently that methods for obtaining this gas, bringing it to the surface, and putting it to use were developed.

Before there was an understanding of what natural gas was, it posed somewhat of a mystery to man. Sometimes, such things as lightning strikes would ignite natural gas that was escaping from under the earth's crust. This would create a fire coming from the earth, burning the natural gas as it seeped out from underground. These fires puzzled most early civilizations, and were the root of much myth and superstition. One of the most famous of these types of flames was found in ancient Greece, on Mount Parnassus approximately 1000 B.C. A goat herdsman came across what looked like a 'burning spring', a flame rising from a fissure in the rock. The Greeks, believing it to be of divine origin, built a temple on the flame. This temple housed a priestess who was known as the Oracle of Delphi, giving out prophecies she claimed were inspired by the flame.

These types of springs became prominent in the religions of India, Greece, and Persia. Unable to explain where these fires came from, they were often regarded as divine, or supernatural. It wasn't until about 500 B.C. that the Chinese discovered the potential to use these fires to their advantage. Finding places where gas was seeping to the surface, the Chinese formed crude pipelines out of bamboo shoots to transport the gas, where it was used to boil sea water, separating the salt and making it drinkable.

Britain was the first country to commercialize the use of natural gas. Around 1785, natural gas produced from coal was used to light houses, as well as streetlights. Manufactured natural gas of this type (as opposed to naturally occurring gas) was first brought to the United States in 1816, when it was used to light the streets of Baltimore, Maryland. However, this manufactured gas was much less efficient, and less environmentally friendly, than modern natural gas that comes from underground.

Naturally occurring natural gas was discovered and identified in America as early as 1626, when French explorers discovered natives igniting gases that were seeping into and around Lake Erie. The American natural gas industry got its beginnings in this area. In 1859, Colonel Edwin Drake (a former railroad conductor who adopted the title 'Colonel' to impress the townspeople) dug the first well. Drake hit oil and natural gas at 69 feet below the surface of the earth.

Most in the industry characterize this well as the beginning of the natural gas industry in America. A two-inch diameter pipeline was built, running 5 and ½ miles from the well to the village of Titusville, Pennsylvania. The construction of this pipeline proved that natural gas could be brought safely and relatively easily from its underground source to be used for practical purposes.

In 1821, the first well specifically intended to obtain natural gas was dug in Fredonia, New York by William Hart. After noticing gas bubbles rising to the surface of a creek, Hart dug a 27-foot well to try and obtain a larger flow of gas to the surface. Hart is regarded by many as the 'father of natural gas' in America. Expanding on Hart's work, the Fredonia Gas Light Company was eventually formed, becoming being the first American natural gas company.

During most of the 19th century, natural gas was used almost exclusively as a source of light. Without a pipeline infrastructure, it was difficult to transport the gas very far, or into homes to be used for heating or cooking. Most of the natural gas produced in this era was manufactured from coal, as opposed to transported from a well. Near the end of the 19th century, with the rise of electricity, natural gas lights were converted to electric lights. This led producers of natural gas to look for new uses for their product.

In 1885, Robert Bunsen invented what is now known as the Bunsen burner. He managed to create a device that mixed natural gas with air in the right proportions, creating a flame that could be safely used for cooking and heating. The invention of the Bunsen burner opened up new opportunities for the use of natural gas in America, and throughout the world. The invention of temperature-regulating thermostatic devices allowed for better use of the heating potential of natural gas, allowing the temperature of the flame to be adjusted and monitored.

Without any way to transport it effectively, natural gas discovered pre-WWII was usually just allowed to vent into the atmosphere, or burnt, when found alongside coal and oil, or simply left in the ground when found alone.

One of the first lengthy pipelines was constructed in 1891. This pipeline was 120 miles long, and carried natural gas from wells in central Indiana to the city of Chicago. However, this early pipeline was very rudimentary, and was not very efficient at transporting natural gas. It wasn't until the 1920s that any significant effort was put into building a pipeline infrastructure. After World War II, welding techniques, pipe rolling, and metallurgical advances allowed for the construction of reliable pipelines. This post-war pipeline construction boom lasted well into the '60s, and allowed for the construction of thousands of miles of pipeline in America.

Once the transportation of natural gas was possible, new uses for natural gas were discovered. These included using natural gas to heat homes and operate appliances such as water heaters and oven ranges. Industry began to use natural gas in manufacturing and processing plants. Also, natural gas was used to heat boilers used to generate electricity. The transportation infrastructure had made natural gas easy to obtain, and it was becoming an increasingly popular form of energy.

A Brief History of Regulation

In 1938, the U.S. government first regulated the natural gas industry. At the time, members of the government believed the natural gas industry to be a 'natural monopoly'. Because of the fear of possible abuses, such as charging unreasonably high prices, and given the rising importance of natural gas to all consumers, the Natural Gas Act was passed. This Act imposed regulations and restrictions on the price of natural gas to protect consumers.

In the 1970s and 1980s, a number of gas shortages and price irregularities indicated that a regulated market was not best for consumers or the natural gas industry. Into the 1980s and early 1990s, the industry gradually moved toward less regulation, allowing for healthy competition and market-based prices.

These moves led to a strengthening of the natural gas market, lowering prices for consumers and allowing for a great deal more natural gas to be discovered. Although not as active as the 1990s, the beginning of the 21st Century has brought with it significant regulation concerning gas quality, standards of conduct for interstate pipelines, and price reporting.

Today, the natural gas industry is regulated by the Federal Energy Regulatory Commission (FERC). While FERC does not deal exclusively with natural gas issues, it is the primary rule making body with respect to the regulation of the natural gas industry.

Competition characterizes the natural gas industry as it is known today. The restructuring of the industry, and the move away from strict regulation, has allowed for increased efficiency and technological improvements. Natural gas is now being obtained more efficiently, cheaply, and easily than ever before. However, the search for more natural gas to serve our ever growing demand requires new techniques and knowledge to obtain it from hard-to-reach places.

Today, the natural gas industry has existed in this country for over 100 years, and it continues to grow. Restructuring and the move toward cleaner-burning fuels have created an enormous market for natural gas across the country.

New technologies are continually developed that allow Americans to use natural gas in new and exciting ways. With all of the advantages of natural gas, it is no wonder that it has become the fuel of choice in this country, and throughout the world.

Proceed to learn how much natural gas there is, and where it might be found!

Natural Gas

The process of getting natural gas out of the ground, and to its final destination to be used, is a complicated one. There is a great deal of behind-the-scenes activity that goes into delivering natural gas to your home, even though it takes only the flick of a switch to turn it on. This section provides an overview of the processes that allow the natural gas industry to get their product out of the ground, and transform it into the natural gas that is used in your homes and in industry.

- The Exploration section outlines how natural gas is found, and how companies decide where to drill wells for it.
- The Extraction section focuses on the drilling process, and how natural gas is brought from its underground reservoirs to the surface.
- The Production section discusses what happens once the well is drilled, including the processing of natural gas once it is brought out from underground.
- The Transport section outlines how the natural gas is transported from the wellhead and processing plant, using the extensive network of pipelines throughout North America.
- The Storage section describes the storage of natural gas, how it is accomplished, and why it is necessary.
- The Distribution section focuses on the delivery of natural gas from the major pipelines to the end users, whoever they may be.
- The Marketing section discusses the role that natural gas marketers play in getting the gas from the wellhead to the end user.

Exploration

The practice of locating natural gas and petroleum deposits has been transformed dramatically in the last 20 years with the advent of extremely advanced, ingenious technology. In the early days of the industry, the only way of locating underground petroleum and natural gas deposits was to search for surface evidence of these underground formations. Those searching for natural gas deposits were forced to scour the earth, looking for seepages of oil or gas emitted from underground before they had any clue that there were deposits underneath. However, because such a low proportion of petroleum and natural gas deposits actually seep to the surface, this made for a very inefficient and difficult exploration process. As the demand for fossil fuel energy has increased dramatically over the past years, so has the necessity for more accurate methods of locating these deposits.

Sources of Data

Technology has allowed for a remarkable increase in the success rate of locating natural gas reservoirs. In this section, it will be outlined how geologists and geophysicists use technology and knowledge of the properties of underground natural gas deposits to gather data that can later be interpreted and used to make educated guesses as to where natural gas deposits exist.

However, it must be remembered that the process of exploring for natural gas and petroleum deposits is characteristically an uncertain one, due to the complexity of searching for something that is often thousands of feet below ground.

Geological Surveys

Exploration for natural gas typically begins with geologists examining the surface structure of the earth, and determining areas where it is geologically likely that petroleum or gas deposits might exist. It was discovered in the mid 1800s that 'anticlinal slopes' had a particularly increased chance of containing petroleum or gas deposits. These anticlinal slopes are areas where the earth has folded up on itself, forming the dome shape that is characteristic of a great number of reservoirs.

By surveying and mapping the surface and sub-surface characteristics of a certain area, the geologist can extrapolate which areas are most likely to contain a petroleum or natural gas reservoir. The geologist has many tools at his disposal to do so, from the outcroppings of rocks on the surface or in valleys and gorges, to the geologic information attained from the rock cuttings and samples obtained from the digging of irrigation ditches, water wells, and other oil and gas wells. This information is all combined to allow the geologist to make inferences as to the fluid content, porosity, permeability, age, and formation sequence of the rocks underneath the surface of a particular area. For example, in the picture shown, a geologist may study the outcroppings of rock to gain insight into the geology of the subsurface areas.

Once the geologist has determined an area where it is geologically possible for a natural gas or petroleum formation to exist, further tests can be performed to gain more detailed data about the potential reservoir area. These tests allow for the more accurate mapping of underground formations, most notably those formations that are commonly associated with natural gas and petroleum reservoirs. These tests are commonly performed by a geophysicist, one who uses technology to find and map underground rock formations.

Seismic Exploration

Arguably the biggest breakthrough in petroleum and natural gas exploration came through the use of basic seismology. Seismology refers to the study of how energy, in the form of seismic waves, moves through the Earth's crust and interacts differently with various types of underground formations. In 1855, L. Palmiere developed the first 'seismograph', an instrument used to detect and record earthquakes. This device was able to pick up and record the vibrations of the earth that occur during an earthquake.

However, it wasn't until 1921 that this technology was applied to the petroleum industry and used to help locate underground fossil fuel formations. The basic concept of seismology is quite simple. As the Earth's crust is composed of different layers, each with its own properties, energy (in the form of seismic waves) travelling underground interacts differently with each of these layers. These seismic waves, emitted from a source, will travel through the earth, but also be reflected back toward the source by the different underground layers. Through seismology, geophysicists are able to artificially create vibrations on the surface and record how these vibrations are reflected back to the surface, revealing the properties of the geology beneath.

An analogy that makes intuitive sense is that of bouncing a rubber ball. A rubber ball that is dropped on concrete will bounce in a much different way than a rubber ball dropped on sand. In the same manner, seismic waves sent underground will reflect off dense layers of rock much differently than extremely porous layers of rock, allowing the geologist to infer from seismic data exactly what layers exist underground and at what depth. While the actual use of seismology in practice is quite a bit more complicated and technical, this basic concept still holds.

Onshore Seismology

In practice, using seismology for exploring onshore areas involves artificially creating seismic waves, the reflection of which are then picked up by sensitive pieces of equipment called 'geophones' that are embedded in the ground. The data picked up by these geophones is then transmitted to a seismic recording truck, which records the data for further interpretation by geophysicists and petroleum reservoir engineers. The drawing shows the basic components of a seismic crew. The source of seismic waves (in this case an underground explosion) creates that reflect off the different layers of the Earth, to be picked up by geophones on the surface and relayed to a seismic recording truck to be interpreted and logged.

Although the seismograph was originally developed to measure earthquakes, it was discovered that much the same sort of vibrations and seismic waves could be produced artificially and used to map underground geologic formations. In the early days of seismic exploration, seismic waves were created using dynamite. These carefully planned, small explosions created the requisite seismic waves, which were then picked up by the geophones, generating data to be interpreted by geophysicists, geologists, and petroleum engineers.

Recently, due to environmental concerns and improved technology, it is often no longer necessary to use explosive charges to generate the needed seismic waves. Instead, most seismic crews use non-explosive seismic technology to generate the required data. This non-explosive technology usually consists of a large heavy-wheeled or tracked-vehicle carrying special equipment designed to create a large impact or series of vibrations. These impacts or vibrations create seismic waves similar to those created by dynamite. In the seismic truck shown, the large piston in the middle is used to create vibrations on the surface of the earth, sending seismic waves that are used to generate useful data.

Offshore Seismology

The same sort of process is used in offshore seismic exploration. When exploring for natural gas that may exist thousands of feet below the seabed floor, which may itself be thousands of feet below sea level, a slightly different method of seismic exploration is used. Instead of trucks and geophones, a ship is used to pick up the seismic data and hydrophones are used to pick up seismic waves underwater. These hydrophones are towed behind the ship in various configurations depending on the needs of the geophysicist. Instead of using dynamite or impacts on the seabed floor, the seismic ship uses a large air gun, which releases bursts of compressed air under the water, creating seismic waves that can travel through the Earth's crust and generate the seismic reflections that are necessary.

Magnetometers

In addition to using seismology to gather data concerning the composition of the Earth's crust, the magnetic properties of underground formations can be measured to generate geological and geophysical data. This is accomplished through the use of magnetometers, which are devices that can measure the small differences in the Earth's magnetic field. In the early days of magnetometers, the devices were large and bulky, and only able to survey a small area at a time.

Gravimeters

In addition to using variances in the Earth's magnetic field, geophysicists can also measure and record the difference in the Earth's gravitational field to gain a better understanding of what is underground. Different underground formations and rock types all have a slightly different effect on the gravitational field that surrounds the Earth. By measuring these minute differences with very sensitive equipment, geophysicists are able to analyze underground formations and develop clearer insight into the types of formations that may lie below ground, and whether or not the formations have the potential for containing hydrocarbons like natural gas.

Exploratory Wells

The best way to gain a full understanding of subsurface geology and the potential for natural gas deposits to exist in a given area is to drill an exploratory well. This consists of digging into the Earth's crust to allow geologists to study the composition of the underground rock layers in detail. In addition to looking for natural gas and petroleum deposits by drilling an exploratory well, geologists also examine the drill cuttings and fluids to gain a better understanding of the geologic features of the area. Logging, explained below, is another tool used in developed as well as exploratory wells. Drilling an exploratory well is an expensive, time consuming effort. Therefore, exploratory wells are only drilled in areas where other data has indicated a high probability of petroleum formations.

Logging

Logging refers to performing tests during or after the drilling process to allow geologists and drill operators to monitor the progress of the well drilling and to gain a clearer picture of subsurface formations. There are many different types of logging, in fact; over 100 different logging tests can be performed, but essentially they consist of a variety of tests that illuminate the true composition and characteristics of the different layers of rock that the well passes through. Logging is also essential during the drilling process. Monitoring logs can ensure that the correct drilling equipment is used and that drilling is not continued if unfavorable conditions develop.

It is beyond the scope of this website to get into detail concerning the various types of logging tests that can be performed. Various types of tests include standard, electric, acoustic, radioactivity, density, induction, caliper, directional and nuclear logging, to name but a few. Two of the most prolific and often performed tests include standard logging and electric logging.

Standard logging consists of examining and recording the physical aspects of a well. For example, the drill cuttings (pieces of rock displaced by the drilling of the well) are all examined and recorded, allowing geologists to physically examine the subsurface rock. Also, core samples are taken by lifting a sample of underground rock intact to the surface, allowing the various layers of rock and their thickness to be examined. These cuttings and cores are often examined using powerful microscopes that can magnify the rock up to 2,000 times. This allows the geologist to examine the porosity and fluid content of the subsurface rock, and to gain a better understanding of the earth in which the well is being drilled.

Electric logging consists of lowering a device used to measure the electric resistance of the rock layers in the 'down hole' portion of the well. This is done by running an electric current through the rock formation and measuring the resistance that it encounters along its way. This gives geologists an idea of the fluid content and characteristics. A newer version of electric logging, called induction electric logging, provides much the same types of readings, but is more easily performed and provides data that is more easily interpreted.

The drilling of an exploratory or developing well is the first contact that a geologist or petroleum engineer has with the actual contents of the subsurface geology. Logging, in its many forms, uses this opportunity to gain a fuller understanding of what actually lies beneath the surface. In addition to providing information specific to that particular well, vast archives of historical logs exist for geologists interested in the geologic features of a given or similar area.

Data Interpretation

There are many sources of data and information for the geologist and geophysicist to use in the exploration for hydrocarbons. However, this raw data alone would be useless without careful and methodical interpretation. Much like putting together a puzzle, the geophysicist uses all of the sources of data available to create a model, or educated guess, as to the structure of the layers of rock under the ground. Some techniques, including seismic exploration, lend themselves well to the construction of a hand- or computer-generated visual interpretation of an underground formation. Other sources of data, such as that obtained from core samples or logging, are taken into account by the geologist when determining the subsurface geological structures. Despite the amazing evolution of technology and exploration techniques, the only way of being sure that a petroleum or natural gas reservoir exists is to drill an exploratory well. Geologists and geophysicists can make their best guesses as to the location of reservoirs, but these are not infallible.

2-D Seismic Interpretation

Two-dimensional seismic imaging refers to geophysicists using the data collected from seismic exploration activities to develop a cross-sectional picture of the underground rock formations. The geophysicist interprets the seismic data obtained from the field, taking the vibration recordings of the seismograph and using them to develop a conceptual model of the composition and thickness of the various layers of rock underground. This process is normally used to map underground formations, and to make estimates based on the geologic structures to determine where it is likely that deposits may exist.

Another technique using basic seismic data is known as 'direct detection.' In the mid-1970s, it was discovered that white bands, called 'bright spots', often appeared on seismic recording strips. These white bands could indicate deposits of hydrocarbons. The nature of porous rock that contains natural gas could often result in reflecting stronger seismic reflections than normal, water-filled rock. Therefore, in these circumstances, the actual natural gas reservoir could be detected directly from the seismic data. However, this does not hold universally. Many of these 'bright spots' do not contain hydrocarbons, and many deposits of hydrocarbons are not indicated by white strips on the seismic data. Therefore, although adding a new technique of locating petroleum and natural gas reservoirs, direct detection is not a completely reliable method.

Computer Assisted Exploration

One of the greatest innovations in the history of petroleum exploration is the use of computers to compile and assemble geologic data into a coherent 'map' of the underground. Use of this computer technology is referred to as 'CAEX', which is short for 'computer assisted exploration'.

With the development of the microprocessor, it has become relatively easy to use computers to assemble seismic data that is collected from the field. This allows for the processing of very large amounts of data, increasing the reliability and informational content of the seismic model. There are three main types of computer-assisted exploration models: two-dimensional (2-D), three-dimensional (3-D), and most recently, four-dimensional (4-D).

These imaging techniques, while relying mainly on seismic data acquired in the field, are becoming more and more sophisticated. Computer technology has advanced so far that it is now possible to incorporate the data obtained from different types of tests, such as logging, production information, and gravimetric testing, which can all be combined to create a 'visualization' of the underground formation. Thus geologists and geophysicists are able to combine all of their sources of data to compile one clear, complete image of subsurface geology. An example of this is shown where a geologist uses an interactive computer generated visualization of 3-D seismic data to explore the subsurface layers.

3-D Seismic Imaging

One of the biggest breakthroughs in computer-aided exploration was the development of three-dimensional (3-D) seismic imaging. Three-D imaging utilizes seismic field data to generate a three dimensional 'picture' of underground formations and geologic features. This, in essence, allows the geophysicist and geologist to see a clear picture of the composition of the Earth's crust in a particular area. This is tremendously useful in allowing for the exploration of petroleum and natural gas, as an actual image could be used to estimate the probability of formations existing in a particular area, and the characteristics of that potential formation. This technology has been extremely successful in raising the success rate of exploration efforts. In fact, using 3-D seismic has been estimated to increase the likelihood of successful reservoir location by 50 percent.

Although this technology is very useful, it is also very costly. Three-D seismic imaging can cost hundreds of thousands of dollars per square mile. The generation of 3-D images requires data to be collected from several thousand locations, as opposed to 2-D imaging, which only requires several hundred data points.

As such, 3-D imaging is a much more involved and prolonged process. Therefore, it is usually used in conjunction with other exploration techniques. For example, a geophysicist may use traditional 2-D modelling and examination of geologic features to determine if there is a probability of the presence of natural gas. Once these basic techniques are used, 3-D seismic imaging may be used only in those areas that have a high probability of containing reservoirs.

In addition to broadly locating petroleum reservoirs, 3-D seismic imaging allows for the more accurate placement of wells to be drilled. This increases the productivity of successful wells, allowing for more petroleum and natural gas to be extracted from the ground. In fact, 3-D seismic can increase the recovery rates of productive wells to 40-50 percent, as opposed to 25-30 percent with traditional 2-D exploration techniques.

In addition to broadly locating petroleum reservoirs, 3-D seismic imaging allows for the more accurate placement of wells to be drilled. This increases the productivity of successful wells, allowing for more petroleum and natural gas to be extracted from the ground. In fact, 3-D seismic can increase the recovery rates of productive wells to 40 to 50 percent or greater, as opposed to 25 to 30 percent with traditional 2-D exploration techniques.

Three-D seismic imaging has become an extremely important tool in the search natural gas. By 1980, only 100 3-D seismic imaging tests had been performed. However, by the mid 1990s, 200 to 300 3-D seismic surveys were being performed each year. In 1996, in the Gulf of Mexico, one of the largest natural gas-producing areas in the U.S., nearly 80 percent of wells drilled in the Gulf were based on 3-D seismic data. In 1993, 75 percent of all onshore exploratory surveys conducted used 3-D seismic imaging.

2-D Seismic Imaging

Two-dimensional (2-D) computer-assisted exploration includes generating an image of subsurface geology much in the same manner as in normal 2-D data interpretation. However, with the aid of computer technology, it is possible to generate more detailed maps more quickly than by the traditional method. In addition, with 2-D CAEX it is possible to use colour graphic displays generated by a computer to highlight geologic features that may not be apparent using traditional 2-D seismic imaging methods.

While 2-D seismic imaging is less complicated and less detailed than 3-D imaging, it must be noted that 3-D imaging techniques were developed prior to 2-D techniques. Thus, although it does not appear to be the logical progression of techniques, the simpler 2-D imaging techniques were actually an extension of 3-D techniques, not the other way around. Because it is simpler, 2-D imaging is much cheaper, and more easily and quickly performed, than 3-D imaging. Because of this,

2-D CAEX imaging may be used in areas that are somewhat likely to contain natural gas deposits, but not likely enough to justify the full cost and time commitment required by 3-D imaging.

4-D Seismic Imaging

One of the latest breakthroughs in seismic exploration and the modelling of underground rock formations has been the introduction of four-dimensional (4-D) seismic imaging. This type of imaging is an extension of 3-D imaging technology. However, instead of achieving a simple, static image of the underground, in 4-D imaging the changes in structures and properties of underground formations are observed over time. Since the fourth dimension in 4-D imaging is time, it is also referred to as 4-D 'time lapse' imaging.

Various seismic readings of a particular area are taken at different times, and this sequence of data is fed into a powerful computer. The different images are amalgamated to create a 'movie' of what is going on under the ground. By studying how seismic images change over time, geologists can gain a better understanding of many properties of the rock, including underground fluid flow, viscosity, temperature and saturation.

Although very important in the exploration process, 4-D seismic images can also be used by petroleum geologists to evaluate the properties of a reservoir, including how it is expected to deplete once petroleum extraction has begun. Using 4-D imaging on a reservoir can increase recovery rates above what can be achieved using 2-D or 3-D imaging. Where the recovery rates using these two types of images are 25 to 30 percent and 40 to 50 percent respectively, the use of 4-D imaging can result in recovery rates of 65 to 70 percent.

Now that we have taken a look at how natural gas deposits are found, the next step in the natural gas line is the process of extraction.

Extraction

Once a potential natural gas deposit has been located by a team of exploration geologists and geophysicists, it is up to a team of drilling experts to dig down to where the natural gas is thought to exist. This section will describe the process of drilling for natural gas, both onshore and offshore.

Although the process of digging deep into the Earth's crust to find deposits of natural gas that may or may not actually exist seems daunting, the industry has developed a number of innovations and techniques that both decrease the cost and increase the efficiency of drilling for natural gas. Advancements in technology have contributed greatly to the increased efficiency and success rate for drilling natural gas wells.

Determining whether to drill a well depends on a variety of factors, including the economic potential of the hoped-for natural gas reservoir. It costs a great deal of money for exploration and production companies to search and drill for natural gas, and there is always the inherent risk that no natural gas will be found.

The exact placement of the drill site depends on many factors, including the nature of the potential formation to be drilled, the characteristics of the subsurface geology, and the depth and size of the target deposit. After the geophysical team identifies the optimal location for a well, it is necessary for the drilling company to ensure that it completes all the necessary steps so that it can legally drill in that area. This usually involves securing permits for the drilling operations, establishment of a legal arrangement to allow the natural gas company to extract and sell the resources under a given area of land, and a design for gathering lines that will connect the well to the pipeline.

There are a variety of potential owners of the land and mineral rights of a given area. To learn more about permitting, leasing, and royalties associated with the extraction of natural gas, visit the regulation section of our website.

If the new well, once drilled, does in fact come in contact with natural gas deposits, it is developed to allow for the extraction of this natural gas, and is termed a 'development' or 'productive' well. At this point, with the well drilled and hydrocarbons present, the well may be completed to facilitate its production of natural gas. However, if the exploration team was incorrect in its estimation of the existence of a marketable quantity of natural gas at a wellsite, the well is termed a 'dry well', and production does not proceed.

Onshore and offshore drilling present unique drilling environments, requiring special techniques and equipment.

Production

Natural gas, as it exists underground, is not exactly the same as the natural gas that comes through the pipelines to our homes and businesses. Natural gas, as we use it, is almost entirely methane. However, when we find natural gas underground, it comes associated with a variety of other trace compounds and gases, as well as oil and water, which must be removed. Natural gas transported through pipelines must meet purity specifications to be allowed in, so most natural gas processing occurs near the well. This section outlines the process of taking raw natural gas from underground formations and processing it into pipeline quality gas, ready for transportation.

Well Completion

Once a natural gas or oil well is drilled, and it has been verified that commercially viable quantities of natural gas are present for extraction, the well must be 'completed' to allow for the flow of petroleum or natural gas out of the formation and up to the surface.

This process includes strengthening the well hole with casing, evaluating the pressure and temperature of the formation, and then installing the proper equipment to ensure an efficient flow of natural gas out of the well.

There are two main types of conventional natural gas wells: natural gas wells and natural gas condensate wells. In addition, there are oil wells that contain "associated" natural gas. In an oil well with associated gas, the natural gas is often used to add pressure to the well and enhance the extraction of the well.

Sometimes associated natural gas exists in large enough quantities to allow its extraction along with the oil. Natural gas specific wells are wells drilled exclusively for natural gas, and contain little or no oil. Shale-gas drilling is an example of wells being drilled for their natural gas resources.

Condensate wells are wells that contain natural gas, as well as a liquid condensate. This condensate is a liquid hydrocarbon mixture that is often separated from the natural gas either at the wellhead, or during the processing of the natural gas. It is important to remember that natural gas, being lighter than air, will naturally rise to the surface of a well. Because of this, in many natural gas and condensate wells, lifting equipment and well treatment are not necessary.

Completing a well consists of a number of steps: installing the well casing, completing the well, installing the wellhead, and installing lifting equipment or treating the formation should that be required. The links below to learn about these aspects of the well completion process:

- Well Casing
- Completion
- The Wellhead
- Lifting and Well Treatment.

Well Casing

Installing well casing is an important part of the drilling and completion process. Well casing consists of a series of metal tubes installed in the freshly drilled hole. Casing strengthens the sides of the well hole, ensures that no oil or natural gas seeps out of the well hole as it is brought to the surface, and keeps other fluids or gases from seeping into the formation through the well.

A good deal of planning is necessary to ensure that the proper casing for each well is installed. The type of casing used depends on the subsurface characteristics of the well, including the diameter of the well and the pressures and temperatures experienced throughout the well. The diameter of the well hole depends on the size of the drill bit used. In most wells, the diameter of the well hole decreases the deeper it is drilled, leading to a type of conical shape that must be taken into account when installing casing.

There are five different types of well casing. They include:

- Conductor Casing
- Surface Casing
- Intermediate Casing
- Liner String
- Production Casing.

Conductor Casing

Conductor casing is installed first, usually prior to the arrival of the drilling rig. The hole for conductor casing is often drilled with a small auger drill, mounted on the back of a truck. Conductor casing is usually no more than 20 to 50 feet long. It is installed to prevent the top of the well from caving in and to help in the process of circulating the drilling fluid up from the bottom of the well. Onshore, this casing is usually 16 to 20 inches in diameter, while offshore casing usually measures 30 to 42 inches. The conductor casing is cemented into place before drilling begins.

Surface Casing

Surface casing is the next type of casing to be installed. It can be anywhere from a few hundred to 2,000 feet long, and is smaller in diameter than the conductor casing. When installed, the surface casing fits inside the top of the conductor casing. The primary purpose of surface casing is to protect fresh water deposits near the surface of the well from being contaminated by leaking hydrocarbons or salt water from deeper underground. It also serves as a conduit for drilling mud returning to the surface, and helps protect the drill hole from being damaged during drilling. Surface casing, like conductor casing, is cemented into place. Regulations often dictate the thickness of the cement to be used to ensure that there is little possibility of freshwater contamination.

Intermediate Casing

Intermediate casing is usually the longest section of casing found in a well. The primary purpose of intermediate casing is to minimize the hazards that come along with subsurface formations that may affect the well. These include abnormal underground pressure zones, underground shale, and formations that might otherwise contaminate the well, such as underground salt-water deposits. In many instances, even though there may be no evidence of an unusual underground formation, intermediate casing is run as insurance against the possibility of such a formation affecting the well. These intermediate casing areas may also be cemented into place for added protection.

Liner Strings

Liner strings are sometimes used instead of intermediate casing. Liner strings are commonly run from the bottom of another type of casing to the open well area. However, liner strings are usually attached to the previous casing with 'hangers', instead of being cemented into place. This type of casing is thus less permanent than intermediate casing.

Production Casing

Production casing, alternatively called the 'oil string' or 'long string,' is installed last and is the deepest section of casing in a well. This is the casing that provides a conduit from the surface of the well to the petroleum-producing formation. The size of the production casing depends on a number of considerations, including the lifting equipment to be used, the number of completions required, and the possibility of deepening the well at a later time. For example, if it is expected that the well will be deepened at a later date, then the production casing must be wide enough to allow the passage of a drill bit later on.

Well casing is a very important part of the completed well. In addition to strengthening the well hole, it provides a conduit to allow hydrocarbons to be extracted without intermingling with other fluids and formations found underground. It is also instrumental in preventing blowouts, allowing the formation to be 'sealed' from the top should dangerous pressure levels be reached. Once the casing has been set, and in most cases cemented into place, proper lifting equipment is installed to bring the hydrocarbons from the formation to the surface. After the casing is installed, tubing is inserted inside the casing, running from the opening well at the top to the formation at the bottom. The hydrocarbons that are extracted go up this tubing to the surface. This tubing may also be attached to pumping systems for more efficient extraction, should that be necessary.

Completion

Well completion commonly refers to the process of finishing a well so that it is ready to produce oil or natural gas. In essence, completion consists of deciding on the characteristics of the intake portion of the well in the targeted hydrocarbon formation. There are a number of types of completions, including:

- Open Hole Completion
- Conventional Perforated Completion
- Sand Exclusion Completion
- Permanent Completion
- Multiple Zone Completion
- Drainhole Completion.

The use of any type of completion depends on the characteristics and location of the hydrocarbon formation to be mined.

Open Hole Completion

Open hole completions are the most basic type and are used in formations that are unlikely to cave in. An open hole completion consists of simply running the casing directly down into the formation, leaving the end of the piping open without any other protective filter. Very often, this type of completion is used on formations that have been 'acidized' or 'fractured.'

Conventional Perforated Completion

Conventional perforated completions consist of production casing being run through the formation. The sides of this casing are perforated, with tiny holes along the sides facing the formation, which allows for the flow of hydrocarbons into the well hole, but still provides a suitable amount of support and protection for the well hole. The process of perforating the casing involves the use of specialized equipment designed to make tiny holes through the casing, cementing, and any other barrier between the formation and the open well. In the past, 'bullet perforators' were used, which were essentially small guns lowered into the well. The guns, when fired from the surface, sent off small bullets that penetrated the casing and cement. Today, 'jet perforating' is preferred. This consists of small, electrically-ignited charges, lowered into the well. When ignited, these charges poke tiny holes through to the formation, in the same manner as bullet perforating.

Sand Exclusion Completion

Sand exclusion completions are designed for production in an area that contains a large amount of loose sand. These completions are designed to allow for the flow of natural gas and oil into the well, but at the same time prevent sand from entering the well. Sand inside the well hole can cause many complications, including erosion of casing and other equipment. The most common methods of keeping sand out of the well hole are screening or filtering systems. These include analyzing the sand experienced in the formation and installing a screen or filter to keep sand particles out. The filter may be either a type of screen hung inside the casing, or a layer of specially-sized gravel outside the casing to filter out the sand. Both types of sand barriers can be used in open holes and perforated completions.

Permanent Completion

Permanent completions are those in which the components are assembled and installed only once. Installing the casing, cementing, perforating, and other completion work is done with small diameter tools to ensure the permanent nature of the completion. Completing a well in this manner can lead to significant cost savings compared to other types.

Multiple Zone Completion

Multiple zone completion is the practice of completing a well so that hydrocarbons from two or more formations may be produced simultaneously, yet separately. For example, a well may be drilled that passes through a number of formations as it descends; alternately, it may be more effective in a horizontal well to add multiple completions to drain the formation efficiently. Although it is common to separate multiple completions so that the fluids from the different formations do not intermingle, the complexity of achieving complete separation can present a barrier. In some instances, the different formations being drilled are close enough to allow fluids to intermingle in the well hole. When it is necessary to prevent this intermingling, hard rubber 'packing' instruments are used to maintain separation among different completions.

Drainhole Completion

Drainhole completions are a form of horizontal or slant drilling. This type of completion consists of drilling out horizontally into the formation from a vertical well, providing a 'drain' for the hydrocarbons to empty into the well. In certain formations, drilling a drainhole completion may allow for more efficient and balanced extraction of the targeted hydrocarbons. Drainhole completions are more commonly associated with oil wells than with natural gas wells.

The Wellhead

The wellhead consists of the pieces of equipment mounted at the opening of the well to manage the extraction of hydrocarbons from the underground formation. It prevents leaking of oil or natural gas out of the well, and also prevents blowouts caused by high pressure. Formations that are under high pressure typically require wellheads that can withstand a great deal of upward pressure from the escaping gases and liquids. These wellheads must be able to withstand pressures of up to 20,000 pounds per square inch (psi). The wellhead consists of three components: the casing head, the tubing head, and the 'christmas tree.'

The casing head consists of heavy fittings that provide a seal between the casing and the surface. The casing head also serves to support the entire length of casing that is run all the way down the well. This piece of equipment typically contains a gripping mechanism that ensures a tight seal between the head and the casing itself.

The tubing head is much like the casing head. It provides a seal between the tubing, which is run inside the casing, and the surface. Like the casing head, the tubing head is designed to support the entire length of the casing, as well as provide connections at the surface, which allow the flow of fluids out of the well to be controlled. The 'christmas tree' is the piece of equipment that fits on top of the casing and tubing heads, and contains tubes and valves that control the flow of hydrocarbons and other fluids out of the well. It commonly contains many branches and is shaped somewhat like a tree, thus its name, 'christmas tree.' The christmas tree is the most visible part of a producing well, and allows for the surface monitoring and regulation of the production of hydrocarbons from a producing well. A typical Christmas tree is about six feet tall.

Lifting and Well Treatment

Once the well is completed, it may begin to produce natural gas. In some instances, the hydrocarbons that exist in pressurized formations will naturally rise up through the well to the surface. This is most commonly the case with natural gas. Since natural gas is lighter than air, once a path to the surface is opened, the pressurized gas will rise to the surface with little or no interference. This is most common for formations containing natural gas alone, or with only a light condensate. In these scenarios, once the christmas tree is installed, the natural gas will flow to the surface without assistance.

In order to more fully understand the nature of the well, a potential test is typically run in the early days of production. This test allows well engineers to determine the maximum amount of natural gas that the well can produce in a 24-hour period. From this and other knowledge of the formation, the engineer may make an estimation on what the 'most efficient recovery rate', or MER will be. The MER is the rate at which the greatest amount of natural gas may be extracted without harming the formation itself.

Another important aspect of producing wells is the 'decline rate'. When a well is first drilled, the formation is under pressure and produces natural gas at a very high rate. However, as more and more natural gas is extracted from the formation, the production rate of the well decreases. This is known as the decline rate. Certain techniques, including lifting and well stimulation, can increase the production rate of a well.

In some natural gas wells, and oil wells that have associated natural gas, it is more difficult to ensure an efficient flow of hydrocarbons up the well. The underground formation may be very 'tight', making the movement of petroleum through the formation and up the well a very slow and inefficient process. In these cases, lifting equipment or well treatment is required. Lifting equipment consists of a variety of specialized equipment used to help 'lift' petroleum out of a formation.

This is most commonly used to extract oil from a formation. Because oil is found as a viscous liquid, it takes some coaxing to extract it from underground. Various types of lifting equipment are available, but the most common lifting method is known as 'rod pumping'. Rod pumping is powered by a surface pump that moves a cable and rod up and down in the well, providing the lifting pressure required to bring the oil to the surface. The most common type of cable rod lifting equipment is the 'horse head' or conventional beam pump.

These pumps are recognizable by the distinctive shape of the cable feeding fixture, which resembles a horse's head.

Well Treatment

Well treatment is another method of ensuring the efficient flow of hydrocarbons out of a formation. Essentially, this type of well stimulation consists of injecting acid, water, or gases into the well to open up the formation and allow the petroleum to flow through the formation more easily. Acidizing a well consists of injecting acid (usually hydrochloric acid) into the well.

In limestone or carbonate formations, the acid dissolves portions of the rock in the formation, opening up existing spaces to allow for the flow of petroleum. Fracturing consists of injecting a fluid into the well, the pressure of which 'cracks' or opens up fractures already present in the formation. In addition to the fluid being injected, 'propping agents' are also used. These propping agents can consist of sand, glass beads, epoxy, or silica sand, and serve to prop open the newly widened fissures in the formation. Hydraulic fracturing involves the injection of water into the formation, while CO2 fracturing uses gaseous carbon dioxide. Fracturing, acidizing, and lifting equipment may all be used on the same well to increase permeability, widening the pores of the formation.

These techniques have been more common to oil wells, but are increasingly being applied to increase the extraction rate for gas wells, particularly hydraulic fracturing. As deeper and less conventional natural gas wells are drilled, it is becoming more common to use stimulation techniques on gas wells.

The next step in the process of producing natural gas is processing. This involves taking the 'raw' natural gas obtained from underground, removing impurities, and ensuring that the gas is ready for use prior to being transported to its destination.

Processing Natural Gas

Natural gas, as it is used by consumers, is much different from the natural gas that is brought from underground up to the wellhead. Although the processing of natural gas is in many respects less complicated than the processing and refining of crude oil, it is equally as necessary before its use by end users.

The natural gas used by consumers is composed almost entirely of methane. However, natural gas found at the wellhead, although still composed primarily of methane, is by no means as pure. Raw natural gas comes from three types of wells: oil wells, gas wells, and condensate wells. Natural gas that comes from oil wells is typically termed 'associated gas'. This gas can exist separate from oil in the formation (free gas), or dissolved in the crude oil (dissolved gas). Natural gas from gas and condensate wells, in which there is little or no crude oil, is termed 'nonassociated gas'. Gas wells typically produce raw natural gas by itself, while condensate wells produce free natural gas along with a semi-liquid hydrocarbon condensate. Whatever the source of the natural gas, once separated from crude oil (if present) it commonly exists in mixtures with other hydrocarbons; principally ethane, propane, butane, and pentanes. In addition, raw natural gas contains water vapour, hydrogen sulfide (H_2S), carbon dioxide, helium, nitrogen, and other compounds.

Natural gas processing consists of separating all of the various hydrocarbons and fluids from the pure natural gas, to produce what is known as 'pipeline quality' dry natural gas. Major transportation pipelines usually impose restrictions on the make-up of the natural gas that is allowed into the pipeline. That means that before the natural gas can be transported it must be purified. While the ethane, propane, butane, and pentanes must be removed from natural gas, this does not mean that they are all 'waste products'.

In fact, associated hydrocarbons, known as 'natural gas liquids' (NGLs) can be very valuable by-products of natural gas processing. NGLs include ethane, propane, butane, iso-butane, and natural gasoline. These NGLs are sold separately and have a variety of different uses; including enhancing oil recovery in oil wells, providing raw materials for oil refineries or petrochemical plants, and as sources of energy.

While some of the needed processing can be accomplished at or near the wellhead (field processing), the complete processing of natural gas takes place at a processing plant, usually located in a natural gas producing region. The extracted natural gas is transported to these processing plants through a network of gathering pipelines, which are small-diameter, low pressure pipes. A complex gathering system can consist of thousands of miles of pipes, interconnecting the processing plant to upwards of 100 wells in the area. According to the American Gas Association's Gas Facts 2000, there was an estimated 36,100 miles of gathering system pipelines in the U.S. in 1999.

In addition to processing done at the wellhead and at centralized processing plants, some final processing is also sometimes accomplished at 'straddle extraction plants'. These plants are located on major pipeline systems. Although the natural gas that arrives at these straddle extraction plants is already of pipeline quality, in certain instances there still exist small quantities of NGLs, which are extracted at the straddle plants.

The actual practice of processing natural gas to pipeline dry gas quality levels can be quite complex, but usually involves four main processes to remove the various impurities:

- Oil and Condensate Removal
- Water Removal
- Separation of Natural Gas Liquids
- Sulfur and Carbon Dioxide Removal.

In addition to the four processes above, heaters and scrubbers are installed, usually at or near the wellhead. The scrubbers serve primarily to remove sand and other large-particle impurities. The heaters ensure that the temperature of the gas does not drop too low. With natural gas that contains even low quantities of water, natural gas hydrates have a tendency to form when temperatures drop.

These hydrates are solid or semi-solid compounds, resembling ice like crystals. Should these hydrates accumulate, they can impede the passage of natural gas through valves and gathering systems. To reduce the occurrence of hydrates, small natural gas-fired heating units are typically installed along the gathering pipe wherever it is likely that hydrates may form.

Oil and Condensate Removal

In order to process and transport associated dissolved natural gas, it must be separated from the oil in which it is dissolved. This separation of natural gas from oil is most often done using equipment installed at or near the wellhead.

The actual process used to separate oil from natural gas, as well as the equipment that is used, can vary widely. Although dry pipeline quality natural gas is virtually identical across different geographic areas, raw natural gas from different regions may have different compositions and separation requirements. In many instances, natural gas is dissolved in oil underground primarily due to the pressure that the formation is under. When this natural gas and oil is produced, it is possible that it will separate on its own, simply due to decreased pressure; much like opening a can of soda pop allows the release of dissolved carbon dioxide. In these cases, separation of oil and gas is relatively easy, and the two hydrocarbons are sent separate ways for further processing. The most basic type of separator is known as a conventional separator. It consists of a simple closed tank, where the force of gravity serves to separate the heavier liquids like oil, and the lighter gases, like natural gas.

In certain instances, however, specialized equipment is necessary to separate oil and natural gas. An example of this type of equipment is the Low-Temperature Separator (LTX). This is most often used for wells producing high pressure gas along with light crude oil or condensate. These separators use pressure differentials to cool the wet natural gas and separate the oil and condensate. Wet gas enters the separator, being cooled slightly by a heat exchanger. The gas then travels through a high pressure liquid 'knockout', which serves to remove any liquids into a low-temperature separator.

The gas then flows into this low-temperature separator through a choke mechanism, which expands the gas as it enters the separator. This rapid expansion of the gas allows for the lowering of the temperature in the separator. After liquid removal, the dry gas then travels back through the heat exchanger and is warmed by the incoming wet gas. By varying the pressure of the gas in various sections of the separator, it is possible to vary the temperature, which causes the oil and some water to be condensed out of the wet gas stream. This basic pressure-temperature relationship can work in reverse as well, to extract gas from a liquid oil stream.

Water Removal

In addition to separating oil and some condensate from the wet gas stream, it is necessary to remove most of the associated water. Most of the liquid, free water associated with extracted natural gas is removed by simple separation methods at or near the wellhead. However, the removal of the water vapour that exists in solution in natural gas requires a more complex treatment. This treatment consists of 'dehydrating' the natural gas, which usually involves one of two processes: either absorption, or adsorption.

Absorption occurs when the water vapour is taken out by a dehydrating agent. Adsorption occurs when the water vapour is condensed and collected on the surface.

Glycol Dehydration

An example of absorption dehydration is known as Glycol Dehydration. In this process, a liquid desiccant dehydrator serves to absorb water vapour from the gas stream. Glycol, the principal agent in this process, has a chemical affinity for water. This means that, when in contact with a stream of natural gas that contains water, glycol will serve to 'steal' the water out of the gas stream. Essentially, glycol dehydration involves using a glycol solution, usually either diethylene glycol (DEG) or triethylene glycol (TEG), which is brought into contact with the wet gas stream in what is called the 'contactor'. The glycol solution will absorb water from the wet gas. Once absorbed, the glycol particles become heavier and sink to the bottom of the contactor where they are removed. The natural gas, having been stripped of most of its water content, is then transported out of the dehydrator. The glycol solution, bearing all of the water stripped from the natural gas, is put through a specialized boiler designed to vaporize only the water out of the solution. While water has a boiling point of 212 degrees Fahrenheit, glycol does not boil until 400 degrees Fahrenheit. This boiling point differential makes it relatively easy to remove water from the glycol solution, allowing it be reused in the dehydration process.

A new innovation in this process has been the addition of flash tank separator-condensers. As well as absorbing water from the wet gas stream, the glycol solution occasionally carries with it small amounts of methane and other compounds found in the wet gas. In the past, this methane was simply vented out of the boiler. In addition to losing a portion of the natural gas that was extracted, this venting contributes to air pollution and the greenhouse effect. In order to decrease the amount of methane and other compounds that are lost, flash tank separator-condensers work to remove these compounds before the glycol solution reaches the boiler.

Essentially, a flash tank separator consists of a device that reduces the pressure of the glycol solution stream, allowing the methane and other hydrocarbons to vaporize ('flash'). The glycol solution then travels to the boiler, which may also be fitted with air or water cooled condensers, which serve to capture any remaining organic compounds that may remain in the glycol solution. In practice, according to the Department of Energy's Office of Fossil Energy, these systems have been shown to recover 90 to 99 percent of methane that would otherwise be flared into the atmosphere.

Solid-Desiccant Dehydration

Solid-desiccant dehydration is the primary form of dehydrating natural gas using adsorption, and usually consists of two or more adsorption towers, which are filled with a solid desiccant. Typical desiccants include activated alumina or a granular silica gel material. Wet natural gas is passed through these towers, from top to bottom. As the wet gas passes around the particles of desiccant material, water is retained on the surface of these desiccant particles. Passing through the entire desiccant bed, almost all of the water is adsorbed onto the desiccant material, leaving the dry gas to exit the bottom of the tower.

Solid-desiccant dehydrators are typically more effective than glycol dehydrators, and are usually installed as a type of straddle system along natural gas pipelines. These types of dehydration systems are best suited for large volumes of gas under very high pressure, and are thus usually located on a pipeline downstream of a compressor station. Two or more towers are required due to the fact that after a certain period of use, the desiccant in a particular tower becomes saturated with water. To 'regenerate' the desiccant, a high-temperature heater is used to heat gas to a very high temperature. Passing this heated gas through a saturated desiccant bed vaporizes the water in the desiccant tower, leaving it dry and allowing for further natural gas dehydration.

Separation of Natural Gas Liquids

Natural gas coming directly from a well contains many natural gas liquids that are commonly removed. In most instances, natural gas liquids (NGLs) have a higher value as separate products, and it is thus economical to remove them from the gas stream. The removal of natural gas liquids usually takes place in a relatively centralized processing plant, and uses techniques similar to those used to dehydrate natural gas.

There are two basic steps to the treatment of natural gas liquids in the natural gas stream. First, the liquids must be extracted from the natural gas. Second, these natural gas liquids must be separated themselves, down to their base components.

NGL Extraction

There are two principle techniques for removing NGLs from the natural gas stream: the absorption method and the cryogenic expander process. According to the Gas Processors Association, these two processes account for around 90 percent of total natural gas liquids production.

The Absorption Method

The absorption method of NGL extraction is very similar to using absorption for dehydration. The main difference is that, in NGL absorption, an absorbing oil is used as opposed to glycol. This absorbing oil has an 'affinity' for NGLs in much the same manner as glycol has an affinity for water. Before the oil has picked up any NGLs, it is termed 'lean' absorption oil. As the natural gas is passed through an absorption tower, it is brought into contact with the absorption oil which soaks up a high proportion of the NGLs. The 'rich' absorption oil, now containing NGLs, exits the absorption tower through the bottom. It is now a mixture of absorption oil, propane, butanes, pentanes, and other heavier hydrocarbons. The rich oil is fed into lean oil stills, where the mixture is heated to a temperature above the boiling point of the NGLs, but below that of the oil.

This process allows for the recovery of around 75 percent of butanes, and 85 - 90 percent of pentanes and heavier molecules from the natural gas stream.

The basic absorption process above can be modified to improve its effectiveness, or to target the extraction of specific NGLs. In the refrigerated oil absorption method, where the lean oil is cooled through refrigeration, propane recovery can be upwards of 90 percent, and around 40 percent of ethane can be extracted from the natural gas stream. Extraction of the other, heavier NGLs can be close to 100 percent using this process.

The Cryogenic Expansion Process

Cryogenic processes are also used to extract NGLs from natural gas. While absorption methods can extract almost all of the heavier NGLs, the lighter hydrocarbons, such as ethane, are often more difficult to recover from the natural gas stream. In certain instances, it is economic to simply leave the lighter NGLs in the natural gas stream. However, if it is economic to extract ethane and other lighter hydrocarbons, cryogenic processes are required for high recovery rates. Essentially, cryogenic processes consist of dropping the temperature of the gas stream to around -120 degrees Fahrenheit.

There are a number of different ways of chilling the gas to these temperatures, but one of the most effective is known as the turbo expander process. In this process, external refrigerants are used to cool the natural gas stream.

Then, an expansion turbine is used to rapidly expand the chilled gases, which causes the temperature to drop significantly. This rapid temperature drop condenses ethane and other hydrocarbons in the gas stream, while maintaining methane in gaseous form.

This process allows for the recovery of about 90 to 95 percent of the ethane originally in the gas stream. In addition, the expansion turbine is able to convert some of the energy released when the natural gas stream is expanded into recompressing the gaseous methane effluent, thus saving energy costs associated with extracting ethane.

The extraction of NGLs from the natural gas stream produces both cleaner, purer natural gas, as well as the valuable hydrocarbons that are the NGLs themselves.

Natural Gas Liquid Fractionation

Once NGLs have been removed from the natural gas stream, they must be broken down into their base components to be useful. That is, the mixed stream of different NGLs must be separated out. The process used to accomplish this task is called fractionation. Fractionation works based on the different boiling points of the different hydrocarbons in the NGL stream. Essentially, fractionation occurs in stages consisting of the boiling off of hydrocarbons one by one. The name of a particular fractionator gives an idea as to its purpose, as it is conventionally named for the hydrocarbon that is boiled off. The entire fractionation process is broken down into steps, starting with the removal of the lighter NGLs from the stream. The particular fractionators are used in the following order:

- Deethanizer - this step separates the ethane from the NGL stream.
- Depropanizer - the next step separates the propane.
- Debutanizer - this step boils off the butanes, leaving the pentanes and heavier hydrocarbons in the NGL stream.
- Butane Splitter or Deisobutanizer - this step separates the iso and normal butanes.

By proceeding from the lightest hydrocarbons to the heaviest, it is possible to separate the different NGLs reasonably easily.

Sulfur and Carbon Dioxide Removal

In addition to water, oil, and NGL removal, one of the most important parts of gas processing involves the removal of sulfur and carbon dioxide. Natural gas from some wells contains significant amounts of sulfur and carbon dioxide. This natural gas, because of the rotten smell provided by its sulfur content, is commonly called 'sour gas'. Sour gas is undesirable because the sulfur compounds it contains can be extremely harmful, even lethal, to breathe.

Sour gas can also be extremely corrosive. In addition, the sulfur that exists in the natural gas stream can be extracted and marketed on its own. In fact, according to the USGS, U.S. sulfur production from gas processing plants accounts for about 15 percent of the total U.S. production of sulfur. For information on the production of sulfur in the United States, visit the USGS here.

Sulfur exists in natural gas as hydrogen sulfide (H_2S), and the gas is usually considered sour if the hydrogen sulfide content exceeds 5.7 milligrams of H_2S per cubic meter of natural gas. The process for removing hydrogen sulfide from sour gas is commonly referred to as 'sweetening' the gas.

The primary process for sweetening sour natural gas is quite similar to the processes of glycol dehydration and NGL absorption. In this case, however, amine solutions are used to remove the hydrogen sulfide. This process is known simply as the 'amine process', or alternatively as the Girdler process, and is used in 95 percent of U.S. gas sweetening operations. The sour gas is run through a tower, which contains the amine solution.

This solution has an affinity for sulfur, and absorbs it much like glycol absorbing water. There are two principle amine solutions used, monoethanolamine (MEA) and diethanolamine (DEA). Either of these compounds, in liquid form, will absorb sulfur compounds from natural gas as it passes through. The effluent gas is virtually free of sulfur compounds, and thus loses its sour gas status. Like the process for NGL extraction and glycol dehydration, the amine solution used can be regenerated (that is, the absorbed sulfur is removed), allowing it to be reused to treat more sour gas. Although most sour gas sweetening involves the amine absorption process, it is also possible to use solid desiccants like iron sponges to remove the sulfide and carbon dioxide.

Sulfur can be sold and used if reduced to its elemental form. Elemental sulfur is a bright yellow powder like material, and can often be seen in large piles near gas treatment plants, as is shown. In order to recover elemental sulfur from the gas processing plant, the sulfur containing discharge from a gas sweetening process must be further treated. The process used to recover sulfur is known as the Claus process, and involves using thermal and catalytic reactions to extract the elemental sulfur from the hydrogen sulfide solution.

In all, the Claus process is usually able to recover 97 percent of the sulfur that has been removed from the natural gas stream. Since it is such a polluting and harmful substance, further filtering, incineration, and 'tail gas' clean up efforts ensure that well over 98 percent of the sulfur is recovered.

Gas processing is an instrumental piece of the natural gas value chain. It is instrumental in ensuring that the natural gas intended for use is as clean and pure as possible, making it the clean burning and environmentally sound energy choice.

Once the natural gas has been fully processed, and is ready to be consumed, it must be transported from those areas that produce natural gas, to those areas that require it.

Peak Oil

Peak oil is the point in time when the maximum rate of global petroleum extraction is reached, after which the rate of production enters terminal decline. This concept is based on the observed production rates of individual oil wells, projected reserves and the combined production rate of a field of related oil wells. The aggregate production rate from an oil field over time usually grows exponentially until the rate peaks and then declines—sometimes rapidly—until the field is depleted. This concept is derived from the Hubbert curve, and has been shown to be applicable to the sum of a nation's domestic production rate, and is similarly applied to the global rate of petroleum production. Peak oil is often confused with oil depletion; peak oil is the point of maximum production while depletion refers to a period of falling reserves and supply.

M. King Hubbert created and first used the models behind peak oil in 1956 to accurately predict that United States oil production would peak between 1965 and 1970. His logistic model, now called Hubbert peak theory, and its variants have described with reasonable accuracy the peak and decline of production from oil wells, fields, regions, and countries, and has also proved useful in other limited-resource production-domains. According to the Hubbert model, the production rate of a limited resource will follow a roughly symmetrical logistic distribution curve (sometimes incorrectly compared to a bell-shaped curve) based on the limits of exploitability and market pressures.

Some observers, such as petroleum industry experts Kenneth S. Deffeyes and Matthew Simmons, predict negative global economy implications following a post-peak production decline—and oil price increase—due to the high dependence of most modern industrial transport, agricultural, and industrial systems on the low cost and high availability of oil. Predictions vary greatly as to what exactly these negative effects would be.

Optimistic estimations of peak production forecast the global decline will begin by 2020 or later, and assume major investments in alternatives will occur before a crisis, without requiring major changes in the lifestyle of heavily oil-consuming nations. These models show the price of oil at first escalating and then retreating as other types of fuel and energy sources are used. Pessimistic predictions of future oil production operate on the thesis that either the peak has already occurred, that oil production is on the cusp of the peak, or that it will occur shortly. The International Energy Agency (IEA) says production of conventional crude oil peaked in 2006. Throughout the first two quarters of 2008, there were signs that a global recession was being made worse by a series of record oil prices.

Demand for Oil

The demand side of peak oil is concerned with the consumption over time, and the growth of this demand. World crude oil demand grew an average of 1.76% per year from 1994 to 2006, with a high of 3.4% in 2003-2004. After reaching a high of 85.6 million barrels (13,610,000 m^3) per day in 2007, world consumption decreased in both 2008 and 2009 by a total of 1.8%, due to rising fuel costs. Despite this lull, world demand for oil is projected to increase 21% over 2007 levels by 2030 (104 million barrels per day (16.5×10^6 m^3/d) from 86 million barrels (13.7×10^6 m^3)), due in large part to increases in demand from the transportation sector. A study published in the journal Energy Policy predicted demand would surpass supply by 2015 (unless constrained by strong recession pressures caused by reduced supply). Energy demand is distributed amongst four broad sectors: transportation, residential, commercial, and industrial.

In terms of oil use, transportation is the largest sector and the one that has seen the largest growth in demand in recent decades. This growth has largely come from new demand for personal-use vehicles powered by internal combustion engines. This sector also has the highest consumption rates, accounting for approximately 68.9% of the oil used in the United States in 2006, and 55% of oil use worldwide as documented in the Hirsch report. Transportation is therefore of particular interest to those seeking to mitigate the effects of peak oil.

Although demand growth is highest in the developing world, the United States is the world's largest consumer of petroleum. Between 1995 and 2005, U.S. consumption grew from 17,700,000 barrels per day (2,810,000 m^3/d) to 20,700,000 barrels per day (3,290,000 m^3/d), a 3,000,000 barrels per day (480,000 m^3/d) increase. China, by comparison, increased consumption from 3,400,000 barrels per day (540,000 m^3/d) to 7,000,000 barrels per day (1,100,000 m^3/d), an increase of 3,600,000 barrels per day (570,000 m^3/d), in the same time frame.

As countries develop, industry and higher living standards drive up energy use, most often of oil. Thriving economies, such as China and India, are quickly becoming large oil consumers. China has seen oil consumption grow by 8% yearly since 2002, doubling from 1996-2006. In 2008, auto sales in China were expected to grow by as much as 15-20%, resulting in part from economic growth rates of over 10%

for five years in a row. Although swift, continued growth in China is often predicted, others predict China's export-dominated economy will not continue such growth trends due to wage and price inflation and reduced demand from the United States. India's oil imports are expected to more than triple from 2005 levels by 2020, rising to 5 million barrels per day (790×10^3 m^3/d).

The International Energy Agency estimated in January 2009 that oil demand fell in 2008 by 0.3%, and that it would fall by 0.6% in 2009. Oil consumption had not fallen for two years in a row since 1982-1983.

The Energy Information Administration (EIA) estimated that the United States' demand for petroleum-based transportation fuels fell 7.1% in 2008, which is "the steepest one-year decline since at least 1950." The agency stated that gasoline usage in the United States may have peaked in 2007, in part due to increasing interest in and mandates for use of biofuels and energy efficiency.

The EIA now expects global oil demand to increase by about 1,600,000 barrels per day (250,000 m^3/d) in 2010. Asian economies, in particular China, will lead the increase. China's oil demand may rise more than 5% compared with a 3.7% gain in 2009, the CNPC said.

Population

Another significant factor on petroleum demand has been human population growth. Oil production per capita peaked in 1979. The United States Census Bureau predicts that the world population in 2030 will be almost double that of 1980. Author Matt Savinar predicts that oil production in 2030 will have declined back to 1980 levels as worldwide demand for oil significantly out-paces production. Physicist Albert Bartlett argues that the decline of the rate of oil production per capita has gone undiscussed because population control is considered politically incorrect by some.

Oil production per capita has declined from 5.26 barrels per year (0.836 m^3/a) in 1980 to 4.44 barrels per year (0.706 m^3/a) in 1993, but then increased to 4.79 barrels per year (0.762 m^3/a) in 2005. In 2006, the world oil production took a downturn from 84.631 to 84.597 million barrels per day (13.4553×10^6 to 13.4498×10^6 m^3/d) although population has continued to increase. This has caused the oil production per capita to drop again to 4.73 barrels per year (0.752 m^3/a).

One factor that has so far helped ameliorate the effect of population growth on demand is the decline of population growth rate since the 1970s. In 1970, the population grew at 2.1%. By 2007, the growth rate had declined to 1.167%. However, oil production was, until 2005, still outpacing population growth to meet demand. World population grew by 6.2% from 6.07 billion in 2000 to 6.45 billion in 2005, whereas according to BP, global oil production during that same period increased from 74.9 to 81.1 million barrels (11.91×10^6 to 12.89×10^6 m^3), or by 8.2%. or according to EIA, from 77.762 to 84.631 million barrels (12.3632×10^6 to 13.4553×10^6 m^3), or by 8.8%.

Agricultural Effects and Population Limits

Since supplies of oil and gas are essential to modern agriculture techniques, a fall in global oil supplies could cause spiking food prices and unprecedented famine in the coming decades. Geologist Dale Allen Pfeiffer contends that current population levels are unsustainable, and that to achieve a sustainable economy and avert disaster the United States population would have to be reduced by at least one-third, and world population by two-thirds.

The largest consumer of fossil fuels in modern agriculture is ammonia production (for fertilizer) via the Haber process, which is essential to high-yielding intensive agriculture. The specific fossil fuel input to fertilizer production is primarily natural gas, to provide hydrogen via steam reforming. Given sufficient supplies of renewable electricity, hydrogen can be generated without fossil fuels using methods such as electrolysis. For example, the Vemork hydroelectric plant in Norway used its surplus electricity output to generate renewable ammonia from 1911 to 1971.

Iceland currently generates ammonia using the electrical output from its hydroelectric and geothermal power plants, because Iceland has those resources in abundance while having no domestic hydrocarbon resources, and a high cost for importing natural gas.

Nationalization of Oil Supplies

Another factor affecting global oil supply is the nationalization of oil reserves by producing nations. The nationalization of oil occurs as countries begin to deprivatize oil production and withhold exports. Kate Dourian, Platts' Middle East editor, points out that while estimates of oil reserves may vary, politics have now entered the equation of oil supply. "Some countries are becoming off limits. Major oil companies operating in Venezuela find themselves in a difficult position because of the growing nationalization of that resource. These countries are now reluctant to share their reserves." According to consulting firm PFC Energy, only 7% of the world's estimated oil and gas reserves are in countries that allow companies like ExxonMobil free rein. Fully 65% are in the hands of state-owned companies such as Saudi Aramco, with the rest in countries such as Russia and Venezuela, where access by Western European and North American companies is difficult.

The PFC study implies political factors are limiting capacity increases in Mexico, Venezuela, Iran, Iraq, Kuwait, and Russia. Saudi Arabia is also limiting capacity expansion, but because of a self-imposed cap, unlike the other countries.

As a result of not having access to countries amenable to oil exploration, ExxonMobil is not making nearly the investment in finding new oil that it did in 1981.

Cartel Influence on Supply

OPEC is an alliance between 12 diverse oil producing countries (Algeria, Angola, Ecuador, Iran, Iraq, Kuwait, Libya, Nigeria, Qatar, Saudi Arabia, the United Arab Emirates, and Venezuela) to control the supply of oil. OPEC's power was consolidated as various countries nationalized their oil holdings, and wrested decision-making away from the "Seven Sisters," (Anglo-Iranian, Socony-Vacuum, Royal Dutch Shell, Gulf, Esso, Texaco, and Socal) and created their own oil companies to control the oil. OPEC tries to influence prices by restricting production.

It does this by allocating each member country a quota for production. All 12 members agree to keep prices high by producing at lower levels than they otherwise would. There is no way to verify adherence to the quota, so every member faces the same incentive to 'cheat' the cartel. Washington kept the oil flowing and gained favourable OPEC policies mainly by arming, and propping up Saudi regimes. According to some, the purpose for the second Iraq war is to break the back of OPEC and return control of the oil fields to western oil companies.

Alternately, commodities trader Raymond Learsy, author of *Over a Barrel: Breaking the Middle East Oil Cartel*, contends that OPEC has trained consumers to believe that oil is a much more finite resource than it is. To back his argument, he points to past false alarms and apparent collaboration. He also believes that peak oil analysts are conspiring with OPEC and the oil companies to create a "fabricated drama of peak oil" to drive up oil prices and profits.

It is worth noting oil had risen to a little over $30/barrel at that time. A counter-argument was given in the Huffington Post after he and Steve Andrews, co-founder of ASPO, debated on CNBC in June 2007.

Timing of Peak Oil

In Feb 2010 the US Joint Forces Command issued the Joint Operating Environment 2010 warning US military commands "By 2012, surplus oil production capacity could entirely disappear, and as early as 2015, the shortfall in output could reach nearly 10 million barrels per day."

> *"A severe energy crunch is inevitable without a massive expansion of production and refining capacity. While it is difficult to predict precisely what economic, political, and strategic effects such a shortfall might produce, it surely would reduce the prospects for growth in both the developing and developed worlds. Such an economic slowdown would exacerbate other unresolved tensions, push fragile and failing states further down the path toward collapse, and perhaps have serious economic impact on both China and India.*
>
> *At best, it would lead to periods of harsh economic adjustment. To what extent conservation measures, investments in alternative energy production, and efforts to expand petroleum production from tar sands and shale would mitigate such a period of adjustment is difficult to predict. One should not forget that the Great Depression spawned a number of totalitarian regimes that sought economic prosperity for their nations by ruthless conquest."*
>
> *"Energy production and distribution infrastructure must see significant new investment if energy demand is to be satisfied at a cost compatible with economic growth and prosperity."*
>
> *"The discovery rate for new petroleum and gas fields over the past two decades (with the possible exception of Brazil) provides little reason for optimism that future efforts will find major new fields."*

The military warning of timing issue is graphically noted in the International Energy Agency's (IEA) World Energy Outlook 2010. The IEA graph notes that depleting conventional oil will be replaced by "fields yet to be found" and "fields yet to be developed."

M. King Hubbert initially predicted in 1974 that peak oil would occur in 1995 "if current trends continue." However, in the late 1970s and early 1980s, global oil consumption actually dropped (due to the shift to energy-efficient cars, the shift to electricity and natural gas for heating, and other factors), then rebounded to a lower level of growth in the mid 1980s.

Thus oil production did not peak in 1995, and has climbed to more than double the rate initially projected. This underscores the fact that the only reliable way to identify the timing of peak oil will be in retrospect.

However, predictions have been refined through the years as up-to-date information becomes more readily available, such as new reserve growth data. Predictions of the timing of peak oil include the possibilities that it has recently occurred, that it will occur shortly, or that a plateau of oil production will sustain supply for up to 100 years. None of these predictions dispute the peaking of oil production, but disagree only on when it will occur.

According to Matthew Simmons, former Chairman of Simmons & Company International and author of *Twilight in the Desert: The Coming Saudi Oil Shock and the World Economy*, "...peaking is one of these fuzzy events that you only know clearly when you see it through a rear view mirror, and by then an alternate resolution is generally too late."

Possible Consequences of Peak Oil

The wide use of fossil fuels has been one of the most important stimuli of economic growth and prosperity since the industrial revolution, allowing humans to participate in takedown, or the consumption of energy at a greater rate than it is being replaced. Some believe that when oil production decreases, human culture, and modern technological society will be forced to change drastically. The impact of peak oil will depend heavily on the rate of decline and the development and adoption of effective alternatives. If alternatives are not forthcoming, the products produced with oil (including fertilizers, detergents, solvents, adhesives, and most plastics) would become scarce and expensive.

In 2005, the United States Department of Energy published a report titled *Peaking of World Oil Production: Impacts, Mitigation, & Risk Management*. Known as the Hirsch report, it stated, "The peaking of world oil production presents the U.S. and the world with an unprecedented risk management problem. As peaking is approached, liquid fuel prices and price volatility will increase dramatically, and, without timely mitigation, the economic, social, and political costs will be unprecedented. Viable mitigation options exist on both the supply and demand sides, but to have substantial impact, they must be initiated more than a decade in advance of peaking."

High Oil Prices

Historical Oil Prices

The oil price historically was comparatively low until the 1973 oil crisis and the 1979 energy crisis when it increased more than tenfold during that six year timeframe. Even though the oil price dropped significantly in the following years, it has never come back to the previous levels.

After almost fifteen years of relative stability, the oil price began to increase again during the 2000s until it hit historical heights of \$143 per barrel (2007 inflation adjusted dollars) on 30 June 2008. As these prices were well above those that caused the 1973 and 1979 energy crises, they have contributed to fears of an economic recession similar to that of the early 1980s.

These fears were not without a basis, since the high oil prices began having an effect on the economies, as, for example, indicated by gasoline consumption drop of 0.5% in the first two months of 2008 in the United States. compared to a drop of.4% total in 2007.

It is agreed that the main reason for the price spike in 2005-2008 was strong demand pressure. For example, global consumption of oil rose from 30 billion barrels (4.8×10^9 m^3) in 2004 to 31 billion in 2005. The consumption rates were far above new discoveries in the period, which had fallen to only eight billion barrels of new oil reserves in new accumulations in 2004.

Oil price increases were partially fuelled by reports that petroleum production is at or near full capacity. In June 2005, OPEC admitted that they would 'struggle' to pump enough oil to meet pricing pressures for the fourth quarter of that year.

The decline in the U.S. dollar against other significant currencies from 2007 to 2008 is also cited as a significant reason for the oil price increases, as the dollar lost approximately 14% of its value against the Euro from May 2007 to May 2008. Besides supply and demand pressures, at times security related factors may have contributed to increases in prices, including the War on Terror, missile launches in North Korea, the Crisis between Israel and Lebanon, nuclear brinkmanship between the U.S. and Iran, and reports from the U.S. Department of Energy and others showing a decline in petroleum reserves.

Effects of Rising Oil Prices

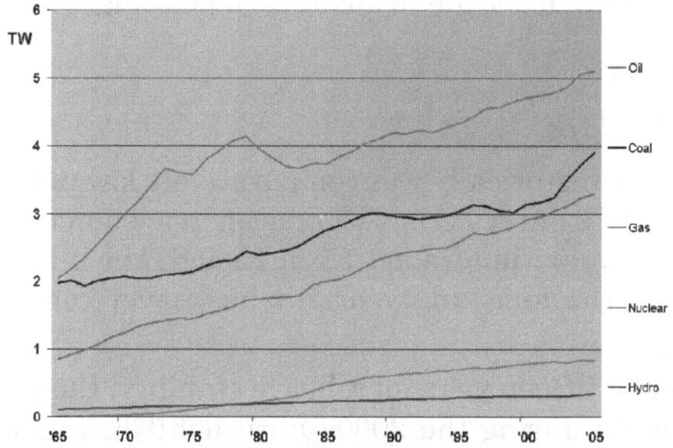

Figure: *World consumption of primary energy by energy type in terawatts (TW), 1965-2005.*

In the past, the price of oil has led to economic recessions, such as the 1973 and 1979 energy crises. The effect the price of oil has on an economy is known as a price shock. In many European countries, which have high taxes on fuels, such price shocks could potentially be mitigated somewhat by temporarily or permanently suspending the taxes as fuel costs rise. This method of softening price shocks is less useful in countries with much lower gas taxes, such as the United States.

Some economists predict that a substitution effect will spur demand for alternate energy sources, such as coal or liquefied natural gas. This substitution can only be temporary, as coal and natural gas are finite resources as well.

Prior to the run-up in fuel prices, many motorists opted for larger, less fuel-efficient sport utility vehicles and full-sized pickups in the United States, Canada, and other countries. This trend has been reversing due to sustained high prices of fuel. The September 2005 sales data for all vehicle vendors indicated SUV sales dropped while small cars sales increased. Hybrid and diesel vehicles are also gaining in popularity.

In 2008, a report by Cambridge Energy Research Associates stated that 2007 had been the year of peak gasoline usage in the United States, and that record energy prices would cause an "enduring shift" in energy consumption practices. According to the report, in April gas consumption had been lower than a year before for the sixth straight month, suggesting 2008 would be the first year U.S. gasoline usage declined in 17 years. The total miles driven in the U.S. peaked in 2006.

The Export Land Model states that after peak oil petroleum exporting countries will be forced to reduce their exports more quickly than their production decreases because of internal demand growth.

Countries that rely on imported petroleum will therefore be affected earlier and more dramatically than exporting countries. Mexico is already in this situation. Internal consumption grew by 5.9% in 2006 in the five biggest exporting countries, and their exports declined by over 3%. It was estimated that by 2010 internal demand would decrease worldwide exports by 2,500,000 barrels per day (400,000 m^3/d).

Canadian economist Jeff Rubin has stated that high oil prices will likely result in increased consumption in developed countries through partial manufacturing de-globalisation of trade. Manufacturing production would move closer to the end consumer to minimise transportation network costs, and therefore a demand decoupling from Gross Domestic Product would occur. Higher oil prices would lead to increased freighting costs and consequently, the manufacturing industry would move back to the developed countries since freight costs would outweigh the current economic wage advantage of developing countries.

Economic research carried out by the International Monetary Fund puts overall price elasticity of demand for oil at -0.025 short term and -0.093 long term.

Long-term Effects on Lifestyle

A majority of Americans live in suburbs, a type of low-density settlement designed around universal personal automobile use. Commentators such as James Howard Kunstler argue that because over 90% of transportation in the U.S. relies on oil, the suburbs' reliance on the automobile is an unsustainable living arrangement. Peak oil would leave many Americans unable to afford petroleum based fuel for their cars, and force them to use bicycles or electric vehicles. Additional options include telecommuting, moving to rural areas, or moving to higher density areas, where walking and public transportation are more viable options.

In the latter two cases, suburbia may become the "slums of the future." The issues of petroleum supply and demand is also a concern for growing cities in developing countries (where urban areas are expected to absorb most of the world's projected 2.3 billion population increase by 2050). Stressing the energy component of future development plans is seen as an important goal.

Methods that have been suggested for mitigating these urban and suburban issues include the use of non-petroleum vehicles such as electric cars, battery electric vehicles, transit-oriented development, Car-free Cities, bicycles, new trains, new pedestrianism, smart growth, shared space, urban consolidation, urban villages, and New Urbanism.

An extensive 2009 report by the United States National Research Council of the Academy of Sciences, commissioned by the United States Congress, stated six main findings.

First, that compact development is likely to reduce "Vehicle Miles Travelled" (VMT) throughout the country.

Second, that doubling residential density in a given area could reduce VMT by as much as 25% if coupled with measures such as increased employment density and improved public transportation.

Third, that higher density, mixed-use developments would produce both direct reductions in CO_2 emissions (from less driving), and indirect reductions (such as from lower amounts of materials used per housing unit, higher efficiency climate control, longer vehicle lifespans, and higher efficiency delivery of goods and services).

Fourth, that although short term reductions in energy use and CO_2 emissions would be modest, that these reductions would grow over time.

Fifth, that a major obstacle to more compact development in the United States is political resistance from local zoning regulators, which would hamper efforts by state and regional governments to participate in land-use planning.

Sixth, the committee agreed that changes in development that would alter driving patterns and building efficiency would have various secondary costs and benefits that are difficult to quantify.

The report recommends that policies supporting compact development (and especially its ability to reduce driving, energy use, and CO_2 emissions) should be encouraged. An economic theory that has been proposed as a remedy is the introduction of a steady state economy. Such a system would include a tax shifting from income to depleting natural resources (and pollution), as well as the limitation of advertising that stimulates demand and population growth. It also includes the institution of policies that move away from globalization and toward localization to conserve energy resources, provide local jobs, and maintain local decision-making authority. Zoning policies would be adjusted to promote resource conservation and eliminate sprawl.

Mitigation

To avoid the serious social and economic implications a global decline in oil production could entail, the 2005 Hirsch report emphasized the need to find alternatives, at least ten to twenty years before the peak, and to phase out the use of petroleum over that time. This was similar to a plan proposed for Sweden that same year. Such mitigation could include energy conservation, fuel substitution, and the use of unconventional oil. Because mitigation can reduce the use of traditional petroleum sources, it can also affect the timing of peak oil and the shape of the Hubbert curve.

Positive Aspects of Peak Oil

Some observers opine that peak oil should be viewed as a positive event. Many such critics reason that if the price of oil rises high enough, the use of alternative clean fuels could help control pollution from fossil fuel use, and mitigate global warming. Permaculture, particularly as expressed in the work of Australian David Holmgren, and others, sees peak oil as holding tremendous potential for positive change, assuming countries act with foresight. The rebuilding of local food networks, energy production, and the general implementation of 'energy descent culture' are argued to be ethical responses to the acknowledgment of finite fossil resources. The Transition Towns movement, started in Totnes, Devon and spread internationally by "The Transition Handbook" (Rob Hopkins), sees the restructuring of society for more local resilience and ecological stewardship as a natural response to the combination of peak oil and climate change.

Criticisms

Some do not agree with peak oil, at least as it has been presented by Matthew Simmons. The president of Royal Dutch Shell's U.S. operations John Hofmeister, while agreeing that conventional oil production will soon start to decline, has criticized Simmons's analysis for being "overly focused on a single country: Saudi Arabia, the world's largest exporter and OPEC swing producer." He also points to the large reserves at the U.S. outer continental shelf, which holds an estimated 100 billion barrels (16×10^9 m^3) of oil and natural gas. As things stand, however, only 15% of those reserves are currently exploitable, a good part of that off the coasts of Louisiana, Alabama, Mississippi, and Texas. Hofmeister also contends that Simmons erred in excluding unconventional sources of oil such as the oil sands of Canada, where Shell is already active.

The Canadian oil sands—a natural combination of sand, water, and oil found largely in Alberta and Saskatchewan—is believed to contain one trillion barrels of oil. Another trillion barrels are also said to be trapped in rocks in Colorado, Utah, and Wyoming, but are in the form of oil shale. These particular reserves present major environmental, social, and economic obstacles to recovery. Hofmeister also claims that if oil companies were allowed to drill more in the United States enough to produce another 2 million barrels per day (320×10^3 m^3/d), oil and gas prices would not be as high as they are in the later part of the 2000 to 2010 decade. He thinks that high energy prices are causing social unrest similar to levels surrounding the Rodney King riots.

Dr. Christoph Rühl, chief economist of BP, repeatedly uttered strong doubts about the peak oil hypothesis:

Physical peak oil, which I have no reason to accept as a valid statement either on theoretical, scientific or ideological grounds, would be insensitive to prices. (...)In fact the whole hypothesis of peak oil – which is that there is a certain amount of oil in the ground, consumed at a certain rate, and then it's finished – does not react to anything.... (Global Warming) is likely to be more of a natural limit than all these peak oil theories combined. (...) Peak oil has been predicted for 150 years. It has never happened, and it will stay this way.

According to Rühl, the main limitations for oil availability are "above ground" and are to be found in the availability of staff, expertise, technology, investment security, money and last but not least in global warming. The oil question is about price and not the basic availability. His views are shared by Daniel Yergin of CERA, who added that the recent high price phase might add to a future demise of the oil industry - not of lack of resources or an apocalyptic shock but the timely and smooth setup of alternatives.

Clive Mather, CEO of Shell Canada, said the Earth's supply of hydrocarbons is almost infinite, referring to hydrocarbons in oil sands. Engineer Peter Huber believes the Canadian oil sands can fuel all of humanity's needs for over 100 years.

Industry blogger Steve Maley echoed some of the points of Yergin, Rühl, Mather and Hofmeister.

Chapter 2

Oil Production Plant

An oil production plant (sometimes called an oil terminal) is a facility which performs processing of production fluids from oil wells in order to separate out key components and prepare them for export. This is distinct from an oil depot, which do not have processing facilities.

Typical production fluids are a mixture of oil, gas and produced water. Many permanent offshore platforms have full oil production facilities onboard. Smaller platforms and subsea wells must export raw production fluid to the nearest production facility, which may be on a nearby offshore processing platform or an onshore terminal.

Process

The production plant is said to begin after the production wing valve on the oil well. The product from each well is piped through the choke valve, which regulates the rate of flow. The flowlines are gathered in a manifold and routed into a separator, which will gravity separate the three components. Once the oil has been separated from the gas and produced water, it is usually routed to a coalescer before being metered and pumped to the onshore terminal.

The produced water is often routed to a hydrocyclone to remove entrained oil and solids and then either re-injected into the reservoir or dumped overboard depending on the circumstances and cleanliness of the water. The associated gas is initially dubbed "wet gas" as it is saturated with water and liquid alkanes. The gas is typically routed through scrubbers, compressors and coolers which will remove the bulk of the liquids. This "dry gas" may be exported, re-injected into the reservoir, used for gas lift, flared or used as fuel for the installation's power generators.

Onshore terminals generally have fired heaters followed by separators and coalescers to stabilise the crude and remove any produced water not separated offshore. Onshore separators tend to operate at a lower pressure than the offshore separators and so more gas is evolved.

The associated gas is generally compressed, dew-pointed and exported via a dedicated pipeline. If gas export is uneconomical then it may be flared. Onshore terminals frequently have large crude oil storage tanks to allow offshore production to continue if the export route becomes unavailable. Export to the refinery is either by pipeline or tanker.

Central Oil Storage

Central oil storage (COS), or central storage, is the term used for a communal heating system pioneered in the middle of the twentieth century.

The concept was simple and effective, using oil (usually kerosene, but sometimes gas oil) in the way that natural gas is used today – it being fed from a central source and metered into individual dwellings on a housing estate. The concept was popular with major oil companies such as Shell and BP. Each householder was obliged to use their oil from their tank, unlike the situation where a consumer owns their own tank and is free to shop around for the best deal.

A typical COS example follows:

Oil would be delivered by road tanker and discharged into a tank capable of holding, say, 5000 gallons. This oil would then find its way into each house, initially through a master pipeline, which then subdivided underground with a branch leading into each property. Each property was provided with a meter located on an outside wall which was read whenever necessary.

A 5000-gallon tank would usually be installed on an estate numbering up to 50 or 60 properties, although they ranged from a low of about 12 up to a high of in the thousands. The majority of tanks were situated above ground in an elevated position. This meant the oil could flow into each house under the influence of gravity. Other tanks were situated underground and were equipped with an electric pump, usually feeding oil to a small 'header' tank, again in an elvated position, to allow gravity to work its magic.

The level of oil in the tank would be checked regularly and new supplies ordered as and when necessary. Therein lies one potential problem with this system – it was possible, through negligence, for the tank to run dry, resulting in a number of properties being left with no heat.

From the early 1980s, their owners began to close down COS sites. The huge increases in the price of oil had led many customers to convert to gas, solid fuel, or even to install their own oil tank in their garden. With fewer and fewer users per site, and maintenance costs remaining the same, the oil companies went through a closure programme, resulting in few sites still being operational in the new millennium.

As at early 2008, the ownership of many sites is unclear. Sites have been sold through a number of companies which have then been closed down, often immediately. This presents problems for householders and local authorities alike, for when a site experiences a problem – such as an unsafe bund wall, or even a leak – there is no one to pursue with a view to getting things put right.

Having visited many sites in South East England, the author believes that such problems will become more common as sites get older and older but receive no maintenance.

Known UK installations

The author worked on a number of such sites in South East England from 1981 to 1983. Locations include :

Hertfordshire : Ashwell, Weston, Letchworth, Hitchin, Dane End, Tring, Berkhamstead, Bovingdon, Boxmoor, Harpenden, Colney Heath, Park Street, Cheshunt x 2 and Hoddesdon x 2. Other known sites include Berkhamstead (Shooters Way),Hitchin (Foster Drive), Stevenage, Hemel Hempstead (Cambrian Way) and Bushey.

Essex : Corringham, Harlow x 2, Stanstead Mountfitchet x 2, and Waltham Abbey. Other known sites include Canvey Island,(700 Houses Supplied off 6 Storage Tanks) Tillingham (Soil Very Corrosive 2nd Only To Sea Bed) All Underground Pipes Relaid 18months after Builders Installed), Great Waltham, Danbury x 2, Ongar, Brentwood, Blackmore, Hullbridge, Bicknacre, Colchester, Eight Ash Green,

Bedfordshire: Luton (many, under the control of Luton Borough Council), Bromham, Leighton Buzzard x 2, Haynes, Kensworth, Harrold, Gt Barford and Cranfield.

Buckinghamshire: Wing, Burnham, Speen.

Berkshire: Newbury.

Cambridgeshire : St Ives, Yaxley, Sawtry x 2, Stilton, Folksworth, Meldreth.

Oxon : Chinnor, Kidlington.

Kent: Crayford.

Leicestershire : Husbands Bosworth.

Hampshire : Winchester, Rogate, Sherfield on Loddon.

Gloucestershire : Eastington, Cirencester, Wootton under Edge.

Suffolk : Haverhill, Roydon (near Diss), Felixstowe, Gorleston, Lowestoft.

Avon : Bristol.

Cambridgeshire : Impington.

Somerset : Evercreech, Midsomer Norton x 4, Paulton

Sussex : (Smallest Known Site 8 Meters The Oasts Ewhurst,) Northiam, Brampton Court Bexhill, Haywards Heath

Norfolk: Attleborough, Gorleston, Harleston.

An external floating roof tank is a storage tank commonly used to store large quantities of petroleum products such as crude oil or condensate. It comprises an open- topped cylindrical steel shell equipped with a roof that floats on the surface of the stored liquid. The roof rises and falls with the liquid level in the tank. As opposed to a fixed roof tank there is no vapour space (ullage) in the floating roof tank (except for very low liquid level situations). In principle, this eliminates breathing losses and greatly reduces the evaporative loss of the stored liquid There is a rim seal system between the tank shell and roof to reduce rim evaporation.

The roof has support legs hanging down into the liquid. At low liquid levels the roof eventually lands and a vapour space forms between the liquid surface and the roof, similar to a fixed roof tank. The support legs are usually retractable to increase the working volume of the tank.

Advantages

External roof tanks are usually installed for environmental or economical reasons to limit product loss and reduce the emission of volatile organic compounds (VOC), an air pollutant.

Normally (roof not landed), there is little vapour space, and consequently a much smaller risk of internal tank explosion.

Disadvantages

Rain water and snow can accumulate on the roof, eventually the roof may sink.

Water on the roof is usually drained from a flexible hose that runs from a drain-sump on the roof, through the stored liquid to a drain valve on the shell at the base of the tank. The hose often develops leaks and drains both water and product.

Note: The following 2 paragraphs (starting "Results of...") are direct wording from a For-Profit company & linked to their website meant for promoting sales of their specific product in a competitive industry where multiple different companies have similar products. This is in direct violation of "Wikipedia is not a...means of promotion," Sections 1) Advocacy, 4) Self-Promotion & 5) Advertising.

There are independent & industry derived testing of this issue and subsequently published papers from these studies that do collaborate the statements made, however they have not been uniformly adopted as fact by the industry. The following paragraphs should be removed as well as the link, and be replaced with documented, independently verifiable testing publications.

Results of Shunt Malfunctions A poorly contacting rim-seal shunt on an EFRT is an example of where sparks might occur during lightning storms. If there is a non-conducting layer on the shell or the shunt, the spark will initially be an air-gap spark that will break down the insulation, followed by current flow in a poorly contacting area resulting in thermal sparking.

Shunt malfunction can result in months or years of downtime, single tank damage, damage to an entire facility, or damage to personnel. Shunts alone are too unreliable to take this chance, as testing has shown that, rather than reducing the risk of fire from lightning strikes, they may actually be increasing risk potential.

Floating Production Storage and Offloading

A floating production, storage and offloading (FPSO) unit is a floating vessel used by the offshore industry for the processing of hydrocarbons and for storage of oil. A FPSO vessel is designed to receive hydrocarbons produced from nearby platforms or subsea template, process them, and store oil until it can be offloaded onto a tanker or transported through a pipeline. FPSOs are preferred in frontier offshore regions as they are easy to install, and do not require a local pipeline infrastructure to export oil. FPSOs can be a conversion of an oil tanker or can be a vessel built specially for the application. A vessel used only to store oil (without processing it) is referred to as a floating storage and offloading vessel (FSO).

History

Oil has been produced from offshore locations since the late 1940s. Originally, all oil platforms sat on the seabed, but as exploration moved to deeper waters and more distant locations in the 1970s, floating production systems came to be used.

The first oil FPSO was the *Shell Castellon*, built in Spain in 1977.

The *Sanha* LPG FPSO operates offshore Angola, and is the first such vessel with complete onboard liquefied petroleum gas processing and export facilities. It can store up to 135,000 cubic meters of LPG while awaiting export tankers for offloading.

In the opposite (discharge and regasification) end of the LNG chain, the first ever conversion of a LNG carrier (Golar LNG owned Moss type LNG carrier) into an LNG floating storage and regasification unit was carried out in 2007 by Keppel shipyard in Singapore. An LNG FPSO works under the same principles an oil FPSO works under, taking the well stream and separating out the natural gas (primarily methane and ethane) and producing LNG, which is stored and offloaded. On July 29, 2009, Shell and Samsung announced an agreement to build up to 10 LNG FPSOs: Already Flex LNG has four contracts for smaller units at the same yard.

On May 20 2011, Royal Dutch Shell announced the planned development of a Floating Liquefied Natural Gas (FLNG) facility, which will be situated 200km off the coast of Western Australia and is due for completion in around 2017. When it is finished, this will be the largest floating offshore facility. It will measure around 488m long and 74m wide, and when fully ballasted will weigh 600,000 tonnes.. It will have a total storage capacity of 436,000 cubic metres of LNG, plus LPG condensate.

Mechanisms

Oil produced from offshore production platforms can be transported to the mainland either by pipeline or by tanker. When a tanker is chosen to transport the oil, it is necessary to accumulate oil in some form of storage tank such that the oil tanker is not continuously occupied during oil production, and is only needed once sufficient oil has been produced to fill the tanker. At this point the transport tanker connects to the stern of the storage unit and offloads oil.

In the early days, the storage units consisted of decommissioned oil tankers, which were stripped down and equipped with process/ production facilities (becoming therefore FPSOs), and were connected to a permanent mooring point. Today, there are two main types of FPSOs, those built converting an existing oil tanker, and those that are purpose-built.

The FPSO design will depend on the area of operation. In benign waters the FPSO may have a simple box shape or it may be a converted tanker. Generally (but not always) the production lines (risers) are connected to a major component of the vessel, called a Turret, which allows the vessel to rotate in order to head into the wind and reduce environmental forces on the moorings. In relatively calm waters, such as in West Africa, turrets can be located externally to the ship structure, hanging off the bow of the FPSO.

For harsher environments like the North Sea, the turret is generally located internally. The turrets and the mooring systems can be designed to be disconnectible or to remain permanently moored. Most ship-shaped FPSOs in the North Sea are purpose-built and are permanently moored.

While most FPSOs are ship-shaped, some FPSOs have a semi-submersible type hull with storage (very rare), or have a cylindrical hull. The inherent symmetry of these FPSO configurations makes turrets unnecessary, so the platforms remain in a fixed orientation.

An FPSO has the capability to carry out some form of separation process. If the unit does not have such facilities, it is generally referred to as a Floating Storage and Offloading unit, and would be operated in conjunction with a production platform. Process plant on FPSO is a core component on facility and forms a key part of production process. Production is usually conducted in 3 phases 1 Separation of Gas. 2 Separation of Water. 3 Separation of oil. Gas recovered/separated during production may be used as fuel on Marine energy resource units (MRU) fitted on board. Gas may be flared off in some cases if MRU is not fitted. Water separation may be carried out using Dehydrators or Hydro Cyclones

Advantages

Floating production, storage and offloading vessels are particularly effective in remote or deepwater locations where seabed pipelines are not cost effective. FPSOs eliminate the need to lay expensive long-distance pipelines from the oil well to an onshore terminal. They can also be used economically in smaller oil fields which can be exhausted in a few years and do not justify the expense of installing a pipeline. Once the field is depleted, the FPSO can be moved to a new location. In areas of the world subject to cyclones (northwestern Australia) or icebergs (Canada), some FPSOs are able to release their mooring/riser turret and steam away to safety in an emergency. The turret sinks beneath the waves and can be reconnected later.

Specific Types

A floating storage and offloading unit (FSO) is a floating storage device, which is a simplified FPSO without the capability for oil or gas processing. Most FSOs are old single hull supertankers that have been converted. An example is *Knock Nevis, ex Seawise Giant*, was for long time world's largest ship, which had been converted to an FSO to be used offshore Qatar.

At the other end of the LNG logistics chain, where the natural gas is brought back to ambient temperature and pressure, ships may also be used as FSRUs. A LNG floating storage and regasification unit (FSRU) is a floating storage and regasification system, which receives liquefied natural gas (LNG) from offloading LNG carriers, and the onboard regasification system provides natural gas send-out through flexible risers and pipeline to shore. Mooring systems for FSO, FPSO & FSU units are available in market which allow the vessel to be moored on a ice sheet. Ice sheet used for mooring this case may be located below th ice sheet under water.

Vessels

Records

The FPSO operating in the deepest water depth is the FPSO BW Pioneer from BW Offshore operated on behalf of Petrobras Americas INC.. The FPSO is moored at a depth of 2,600 m in Block 249 Walker Ridge in the US GoM and is rated for 100,000 bbl/d (16,000 m^3/d). The EPCI contract was awarded in October 2007 and first oil is planned for 3rd Q 2011. The FPSO conversions were done at Keppel Shipyard Tuas in Singapore and the topsides were fabricated in modules at various international vendors. The FPSO has a disconnectable turret (APL). The vessel can disconnect for hurricanes and reconnect with minimal down time. The world's largest FPSO is the *Kizomba A*, with a storage capacity of 2.2 million barrels (350,000 m^3). Built at a cost of over US$800 million by Hyundai Heavy Industries in Ulsan, Korea, it is operated by Esso Exploration Angola (ExxonMobil). Located in 1200 meters (3,940 ft) of water at Deepwater block 200 statute miles (320 km) offshore in the Atlantic Ocean from Angola, Central Africa, it weighs 81,000 tonnes and is 285 meters long, 63 meters wide, and 32 meters high (935 ft by 207 ft (63 m) by 105 ft).

The first FSO in the Gulf of Mexico, The FSO Ta'Kuntah, has been in operation since August 1998. The FSO is under a service agreement with PEMEX Exploration and Production. MODEC owns and operates the FSO. The FSO Ta'Kuntah was installed as part of the Cantarell Field Development in the Gulf of Mexico. The field is located in the Bay of Campeche, offshore Mexico's Yucatan peninsula. The FSO Ta'Kuntah is a converted ULCC tanker with a SOFEC external turret mooring system, two flexible risers connected in a lazy-S configuration between the turret and a pipeline end manifold (PLEM) on the seabed, and a unique offloading system. The FSO is designed to handle 800,000 bbl/d (130,000 m^3/d) with no allowance for downtime.

The world's smallest FPSO is the Crystal Ocean, operating in 137 m of water in the Bass Strait between Australia and Tasmania on the Basker Manta Field. It is leased by Roc Oil (Sydney-based international petroleum exploration and production company) from Rubicon Offshore and is operated on their behalf by AGR Asia Pacific; it is currently producing 5,000 bbl/d (790 m^3/d).

The FPSO in the shallowest water depth of just 13 m is the Armada Perkasa in the Okoro field in Nigeria, West Africa, for Afren Energy. This spread moored (fixed orientation) vessel uses 100 mm, 150 mm and 200 mm bore DeepFlex non-steel flexible risers in a double lazy wave formation (with weights and distributed buoyancy) to accommodate the large motion offsets in an environment of extreme waves and currents.

The Skarv FPSO, developed and engineered by Aker Solutions for BP Norge, will be the most advanced and largest FPSO deployed in the Norwegian Sea, offshore Mid Norway. Skarv is a gas condensate and oil field development. The development will tie in five sub-sea templates, and the FPSO has capacity to include several smaller wells nearby in the future. The process plant on the vessel can handle about 19,000,000 cubic metres per day (670,000,000 cu ft/d) of gas and 13,500 cubic metres per day (480,000 cu ft/d).

An 80 km gas export pipe will tie in to Åsgard transport system. Aker Solutions (formerly Aker Kvaerner) developed the front-end design for the new floating production facility as well as the overall system design for the field and preparation for procurement and project management of the total field development. BP also selected Aker Solutions to perform the detail engineering, procurement and construction management assistance (EPcma) for the Skarv field development. The EPcma contract covers detail engineering and procurement work for the FPSO topsides as well as construction management assistance to BP including hull and topside facilities. The production start for the field is scheduled for August 2011. BP awarded the contract for fabrication of the Skarv FPSO hull to Samsung Heavy Industries in South Korea and the Turret contract to SBM. The FPSO has a length of 292m, breadth of 50.6m and is 29m deep and accommodate 100 people in single cabins. The hull will be delivered in January 2010.

In July 2011, Shell announced the world's largest FPSO, scheduled to come on stream in 2017.

Global Strategic Petroleum Reserves

Global strategic petroleum reserves ("GSPR") refer to crude oil inventories (or stockpiles) held by the government of a particular country, as well as private industry, for the purpose of providing economic and national security during an energy crisis. According to the United States Energy Information Administration, approximately 4.1 billion barrels (650,000,000 m^3) of oil are held in strategic reserves, of which 1.4 billion is government-controlled.

The remainder is held by private industry. At the moment the US Strategic Petroleum Reserve is one of the largest strategic reserves, with much of the remainder held by the other 26 members of the International Energy Agency. Other non-IEA countries have begun creating their own strategic petroleum reserves, with China being the largest of these new reserves. Since current consumption levels are neighboring 0.1 billion barrels (16,000,000 m^3) per day, in the case of a dramatic worldwide drop in oil field output as suggested by some peak oil analysts, the strategic petroleum reserves are unlikely to last for more than a few months.

International Energy Agency Reserves

According to a March 2001 agreement, all 28 members of the International Energy Agency must have a strategic petroleum reserve equal to 90 days of prior year's net oil imports for their respective country. Only net-exporter members of the IEA are exempt from the reserve requirement. The exempt countries are Canada, Denmark, Norway, and the United Kingdom. However, Denmark and the UK have both recently created strategic reserves due to their requirements as European Union members.

Forward Commercial Storage Agreements

To allow oil-exporting countries increased flexibility in their production quotas, there has been an increased movement towards forward commercial storage agreements. These agreements allow petroleum to be stored at an oil-importing country, however the reserves are technically under the control of the oil-exporting country. Oil importing countries benefit from the close access to the commercial reserves, while reducing the costs of access.

Emergency Oil Sharing Agreements

In addition to maintaining a domestic stockpile of petroleum, several countries also have agreements to share their stockpiles in the event of an emergency.

The Japan, New Zealand and South Korea Agreement

In mid-2007 Japan announced a program to share its strategic reserve with other countries in its region. Negotiations are currently underway with New Zealand on an emergency oil-sharing program whereby Japan would make available for purchase its strategic reserves. In an emergency New Zealand would pay the market price plus negotiated option fees for the amount of oil previously held for them by Japan. South Korea and Japan have also agreed to share their oil reserves in case of an emergency.

The United States and Israel Agreement

According to the 1975 Second Sinai withdrawal document signed by the United States and Israel, in an emergency the U.S. is obligated to make oil available for sale to Israel for up to 5 years.

The France, Germany, and Italy agreement

France, Germany and Italy have an oil-sharing agreement to allow each other to purchase their strategic reserves in the event of an emergency.

Africa

South Africa has an SPR. It is managed by PetroSA and the primary facility is the Saldanha Bay oil storage facility, which is a major transit point for oil shipping. Saldanha Bay's six in-ground concrete storage tanks give the facility a storage capacity of 45,000,000 barrels (7,200,000 m^3).

Malawi is considering creating a 21-day reserve of fuel, which is an expansion from the current five day reserve. The government has begun planning for storage facilities in the provinces of Chipoka and Mchinji as well as Kamuzu International Airport.

Kenya is setting up a Strategic Fuel Reserve, similar to that of cereals. The strategic stocks would be procured by the National Oil

Corporation of Kenya and stored by the Kenya Pipeline Company Limited.

Asia

China: In 2007 China announced an expansion of their crude reserves into a two part system. Chinese reserves would consist of a government-controlled strategic reserve complemented by mandated commercial reserves. The government-controlled reserves are being completed in three phases. Phase one consisted of a 101,900,000 barrels (16,200,000 m^3) reserve, mostly completed by the end of 2008.

The second phase of the government-controlled reserves with an additional 170,000,000 barrels (27,000,000 m^3) will be completed by 2011. Recently, Zhang Guobao the head of the National Energy Administration also stated that there will be a third phase that will expand reserves by 204,000,000 barrels (32,400,000 m^3) with the goal of increasing China's SPR to 90 days of supply by 2020.

The planned state reserves of 475,900,000 barrels (75,660,000 m^3) plus the planned enterprise reserves of 209,440,000 barrels (33,298,000 m^3) will provide around 90 days of consumption or a total of 684,340,000 barrels (108,801,000 m^3).

India: India has begun the development of a strategic crude oil reserve sized at 37,400,000 barrels (5,950,000 m^3), enough for two weeks of consumption. Petroleum stocks have been transferred from the Indian Oil Corporation (IndianOil) to the Oil Industry Development Board (OIDB). The OIDB then created the Indian Strategic Petroleum Reserves Ltd (ISPRL) to serve as the controlling government agency for the strategic reserve.

The facilities are:

- Mangalore, State of Karnataka. Capacity of 11.22 million barrels (1,784,000 m^3).
- Padur village, Udipi in the state of Karnataka. Capacity of 18.7 million barrels (2,970,000 m^3).
- Visakhapatnam, State of Andhra Pradesh. Capacity of 7.48 million barrels (1,189,000 m^3).

Japan: As of 2010 Japan has an SPR composed of the following three types of stockpiles:

- State controlled reserves of petroleum at eleven different locations totaling 324,000,000 barrels (51,500,000 m^3). *All numbers, unless otherwise cited, come from p. 177 of this document:*

- o Tomakomai Eastern Oil Reserve Storage Base, 55 storage tanks, total capacity 34 million barrels (5,400,000 m^3).
- o Mutsu-Ogawara Storage Base, 53 storage tanks, total capacity 31 million barrels (4,900,000 m^3).
- o Kuji Storage Base, 3 storage tanks, total capacity 10.5 million barrels (1,670,000 m^3).
- o Akita Storage Base, 15 storage tanks, total capacity 23.4 million barrels (3,720,000 m^3).
- o Fukui Storage Base, 27 storage tanks, total capacity 17.9 million barrels (2,850,000 m^3).
- o Kikuma Underground Petroleum Storage Facility, 8 storage tanks, total capacity 8.9 million barrels (1,410,000 m^3).
- o Shirashima Storage Facility, 8 tankers (4,400,000 barrels (700,000 m^3) each), total capacity 35.2 million barrels (5,600,000 m^3).
- o Kamigotou Storage Base, 7 storage tanks, total capacity 21.45 million barrels (3,410,000 m^3).
- o Kushikino Storage Base, 3 storage tanks, total capacity 10.5 million barrels (1,670,000 m^3).
- o Shibushi Storage Base, 40 storage tanks, total capacity 27.6 million barrels (4,390,000 m^3).
- o Kagoshima former Nippon Oil facility, 4,000,000 barrels (640,000 m^3). This is a forward commercial storage facility with Abu Dhabi.

- Privately held reserves of petroleum held "in accordance with the Petroleum Stockpiling Law" of 129,000,000 barrels (20,500,000 m^3).
- Privately held reserves of petroleum products for another 130,000,000 barrels (21,000,000 m^3).

The state stockpile and the privately held stockpiles total about 583,000,000 barrels (92,700,000 m^3). The Japanese SPR is run by the Japan Oil, Gas and Metals National Corporation.

South Korea

As part of the government's energy security efforts, South Korea holds strategic oil reserves to protect against oil supply disruptions. The country's strategic oil reserve program is managed by the Korea National Oil Corporation, which reports that its system has the capacity to store 116,000,000 barrels (18,400,000 m^3) of oil. As of April 2007, KNOC held 76,000,000 barrels (12,100,000 m^3) of oil in its strategic stockpiles, 64,000,000 barrels (10,200,000 m^3) of crude oil and 12,000,000 barrels (1,900,000 m^3) of petroleum products.

This total amounts to approximately 34 days of net import cover, according to 2006 estimates of demand. KNOC has plans to expand the country's strategic storage capacity from 116,000,000 barrels (18,400,000 m^3) to 146,000,000 barrels (23,200,000 m^3) by 2009, and to fill the emergency reserves to 141,000,000 barrels (22,400,000 m^3) by 2010.

Others

The Philippines has begun plans for a National Petroleum Strategic Reserve by 2010 with an approximate size of 30,000,000 barrels (4,800,000 m^3).

Russia has begun plans for a strategic petroleum reserve. Analysts estimate the size of the Russian SPR would be around 78,000,000 barrels (12,400,000 m^3).

Singapore has an SPR composed of 31.8 million barrels (5,060,000 m^3) of crude oil with an additional 64.5 million barrels (10,250,000 m^3) of oil products for a total of 96,300,000 barrels (15,310,000 m^3).

Taiwan has an SPR with a 1999 reported size of 13,000,000 barrels (2,100,000 m^3). Taiwan's refiners (Kaohsiung 270,000 bbl/d (43,000 m^3/d); Ta-Lin 300,000 bbl/d (48,000 m^3/d); Tao-Yuan 200,000 bbl/d (32,000 m^3/d); Mailiao 150,000 bbl/d) are also required to store at least 30 days of petroleum stocks. As of 2005, these mandated commercial reserves total 27,600,000 barrels (4,390,000 m^3) of strategic petroleum stocks.

Thailand has increased the size of its SPR from 60 days to 70 days of consumption in 2006.

Pakistan has begun plans for a 20 day emergency reserve.

European Union

In the European Union, according to Council Directive 68/414/EEC of 20 December 1968, all 27 members must have a strategic petroleum reserve within the territory of the E.U. equal to at least 90 days average daily internal consumption.

The Czech Republic has a four tank SPR facility in Nelahozeves run by the company CR Mero. The Czech SPR is equal to 100 days of consumption or 20,300,000 barrels (3,230,000 m^3).

Denmark has a reserve of 81 days of consumption, equal to about 1,4 million tonnes of oil products. Not counting reserves held by the military defence.

Finland has an SPR with an approximate size of 62,400,000 barrels (9,920,000 m^3).

France has an SPR with an approximate size of 65,000,000 barrels (10,300,000 m^3).

As of 2000 jet fuel stocks were required for at least 55 days of consumption, with half of those stocks controlled by the *Société Anonyme de Gestion des Stocks de Sécurité* (SAGESS) and the other half controlled by producers.

Germany created the Federal Oil Reserve in 1970, stored in the Etzel salt caverns near Wilhelmshaven in northern Germany, with an initial size of 70 million barrels (11,000,000 m^3). The current German Federal Oil Reserve and the Erdölbevorratungsverband (EBV) (the German stockholding company) mandates that refiners must keep 90 days of stock on hand, giving Germany an approximate reserve size of 250,000,000 barrels (40,000,000 m^3) as of 1997. The German SPR is the largest in Europe.

Hungary has an SPR with approximately 90 days of consumption or 11,880,000 barrels (1,889,000 m^3).

Ireland has approximately 31 days of oil stocks in Ireland and another 9 days of oil stocks held in fellow EU members states. Additionally, they have stock tickets (contracts with a 3rd party where the government has the option to purchase in the event of an emergency) and stocks held by large industry or large consumers. On average Ireland has approximately 100 days of oil available.

Poland has an SPR with approximately 70 days of consumption. Another facility holding 20 additional days of consumption is scheduled to be completed in 2008. Poland also requires oil companies to maintain reserves sufficient for 73 days of production.

Portugal has an SPR with an approximate size of 22,440,000 barrels (3,568,000 m^3).

Slovakia has an SPR with an approximate size of 748,000 barrels (118,900 m^3).

Spain has an SPR with an approximate size of 120,000,000 barrels (19,000,000 m^3).

Sweden has an SPR with an approximate size of 13,290,000 barrels (2,113,000 m^3).

The United Kingdom has created a strategic reserve, the size is unknown.

Russia

Russia has begun accumulating strategic reserves of refined products which will be held by Rosneftegaz, the state-owned company. The reserves will be held at commercial refineries, Transneft facilities, and state reserve facilities.

The current planned size is 14,665,982 barrels (2,331,704.8 m^3).

Switzerland

Switzerland has SPRs consisting of gas, diesel, jet fuel and heating oil for 4.5 months of consumption. The reserves were created in the 1940s and were used for the first time in 2005 following Hurricane Katrina.

Middle East

Iran: In April 2006 the Fars News Agency reported that Iran has begun plans to create an SPR. The National Iranian Oil Company (NIOC) has begun construction of 15 crude oil storage tanks with a planned capacity of 10,000,000 barrels (1,600,000 m^3).

In August 2008, Iran announced plans to expand their SPR with a new facility on Kharg Island containing 4 tanks holding 1,000,000 barrels (160,000 m^3) each. Iran's SPR facilities are:

- Ahwaz. 4 storage tanks, total capacity 2 million barrels (320,000 m^3).
- Omidiyeh. 3 storage tanks, total capacity 3 million barrels (480,000 m^3).
- Goureh. 6 storage tanks, total capacity 4 million barrels (640,000 m^3).
- Sirri Island. 1 storage tank, total capacity 500,000 barrels (79,000 m^3).
- Bahregansar. 1 storage tank, total capacity 500,000 barrels (79,000 m^3).
- Kharg Island. 4 storage tanks, total capacity 4 million barrels (640,000 m^3). Planned facility, not operational yet.

Kuwait: Kuwait has a joint stockpile held in South Korea. The deal gives South Korea first rights to purchase the oil. The current size of the stockpile is 2 million barrels (320,000 m^3).

Others

As of 1975 Israel is believed to have a strategic oil reserve equal to 270 days of consumption.

Jordan has strategic oil reserves equal to 60 days of consumption or 6,240,000 barrels (992,000 m^3).

North America

United States: The United States has the largest reported Strategic Petroleum Reserve with a total capacity of 727,000,000 barrels (115,600,000 m^3). If completely filled, the US SPR could theoretically replace about 60 days of oil imports as the US is estimated to import approximately 12,000,000 barrels per day (1,900,000 m^3/d) of crude oil. According to the US Department of Energy the facilities maximum flow rate is limited to approximately 4,400,000 barrels per day (700,000 m^3/d) when filled to maximum, with flow rate declining as the reserve is drawn down. The U.S. facilities are in salt caverns with locations in:

- Bryan Mound - located near Freeport, Texas. Capacity of 226,000,000 barrels (35,900,000 m^3).
- Big Hill - located near Winnie, Texas. Capacity of 160,000,000 barrels (25,000,000 m^3).
- West Hackberry - located near Lake Charles, Louisiana. Capacity of 219,000,000 barrels (34,800,000 m^3).
- Bayou Choctaw - located near Baton Rouge, Louisiana. Capacity of 72,000,000 barrels (11,400,000 m^3).
- Richton, Mississippi. A planned facility.

The US has also organized the 2-million-barrel (320,000 m^3) Northeast Home Heating Oil Reserve to supply northeast homeowners during shortages.

California

The state of California is considering the creation of Strategic Fuels Reserve.

Oceania

New Zealand has a strategic reserve with a 2008 size of 170,000 tons or 1,200,000 barrels (190,000 m^3). Much of this reserve is based on ticketed option contracts with Australia, Japan, the United Kingdom and the Netherlands, which allow for guaranteed purchases of petroleum in the event of an emergency.

Oil-storage Trade

The oil-storage trade is a trading strategy where oil tank owners and companies that lease storage buy oil for immediate delivery and hold it in their storage tanks, then sell contracts for future delivery at a higher price. When delivery dates approach, they close out existing contracts and sell new ones for future delivery of the same oil. The oil never moves out of storage.

Trading in this fashion is only successful if the forward market is in "contango", that is if the price of oil in the future also known as forward prices are higher than current prices or spot prices. Storing oil became big business in 2008 and 2009, with many participants - including Wall Street giants such as Morgan Stanley, Goldman Sachs or Citicorp - turning sizeable profits simply by sitting on tanks of oil.

It has been estimated that one in twelve of the largest oil tankers are being used for the storage, rather than transportation of oil, and that if lined up end to end, the tankers would stretch out for 26 miles.

Strategic Petroleum Reserve (China)

The Strategic Petroleum Reserve (SPR) is an emergency fuel store of oil maintained by the People's Republic of China National Development and Reform Commission.

Background

In 2007 China announced an expansion of their crude reserves into a two part system. Chinese reserves would consist of a government-controlled strategic reserve complemented by mandated commercial reserves. The government-controlled reserves are being completed in three phases. Phase one consisted of a 101.9 million barrel reserve, mostly completed by the end of 2008. The second phase of the government-controlled reserves with an additional 170 million barrels will be completed by 2011. Recently, Zhang Guobao the head of the National Energy Administration also stated that there will be a third phase that will expand reserves by 204 million barrels with the goal of increasing China's SPR to 90 days of supply by 2020.

Reserve structure

Government Reserves

The government and enterprise reserves are managed by the National Development and Reform Commission(NDRC). Current plans call for government reserves of 475.9 million barrels (101.9 million barrels completed, 374 million barrels planned).

- First phase facilities:
  - o Dalian, Liaoning Province. Capacity of 19 million barrels (3,000,000 m^3), filled as of September 2009.

- Qingdao, Shandong Province. Capacity of 19 million barrels (3,000,000 m^3), filled as of September 2009.
- Zhenhai, Zhejiang Province. 52 storage tanks with a capacity of 33 million barrels (5,200,000 m^3), filled as of December 2007.
- Zhoushan, Zhejiang Province. Capacity of 33 million barrels (5,200,000 m^3), 7.6 million barrels (1,210,000 m^3) filled as of June 2007.

- Second phase facilities:
  - Dushanzi, Karamay City, Xinjiang Uygur Autonomous Region. Planned capacity of 34 million barrels (5,400,000 m^3).
  - Lanzhou, Gansu province. Planned capacity of 19 million barrels (3,000,000 m^3), with completion by 2011.
  - Huangdao, Shandong Province
  - Jinzhou, Liaoning Province
  - Four other facilities, location and size to be determined.
- Third phase facilities:
  - Wanzhou, Chongqing Municipality
  - Henan Province
  - Caofeidian, Hebei Province
  - Tianjin. Planned capacity of 150 million barrels (20,000,000 m^3), with completion by 2023.
- Local government reserves
  - Guangdong Province has begun plans for an expanded reserve from 20 days to 90 days.
  - Hainan Province has begun plans for a reserve.
  - Shanshan, Xinjiang Uygur Autonomous Region has begun plans for a reserve.
  - Caofeidian, Hebei province has begun plans for a reserve.
  - Wanzhou, Chongqing municipality has begun plans for a reserve.

Enterprise Reserves

Currently the enterprise reserves compose a smaller portion of the overall SPR with a 209.44 million barrel strategic reserve planned (35.33 million barrels completed, 51.5 million barrels under construction).

- Commercial oil reserves by major Chinese oil companies, PetroChina, Sinopec and CNOOC
  - o PetroChina facility, Shanshan County, Xinjiang Uygur Autonomous Region. Completed with a capacity of 6.3 million barrels (1,000,000 m^3).
  - o PetroChina facility, Tieling, Liaoning Province. Completed with a capacity of 5.03 million barrels (800,000 m^3), completion October 2008
  - o Sinopec facility, Ningbo, Zhejiang Province. Completed with a capacity of 24 million barrels (3,800,000 m^3).
  - o Sinopec facility, Rizhao, Shandong Province, under construction. Planned capacity of 20.1 million barrels (3,200,000 m^3)
  - o Sinopec facility, Beihai, Guangxi region, under construction. Planned capacity by 2011 of 20.1 million barrels (3,200,000 m^3)
  - o Sinopec facility, Zhanjiang, under construction
  - o Sinopec facility, Caofeidian, under construction
  - o Sinopec facility, Shanghai, under construction
  - o Sinochem facility, under construction
  - o Unknown company, Heilongjiang Province
  - o CNOOC facility, Gansu Province, under construction. Planned capacity of 11.3 million barrels (1,800,000 m^3).
- Oil storage reserves by medium and small Chinese oil companies.

The planned state reserves of 475.9 million barrels plus the planned enterprise reserves of 209.44 million barrels will provide around 90 days of consumption or a total of 685.34 million barrels.

Other SPRs

The U.S. Strategic Petroleum Reserve is the largest emergency supply in the world with the current capacity to hold up to 727 million barrels (115,600,000 m^3). The second largest emergency supply of oil is Japan's with a 2003 reported capacity of 579 million barrels (92,100,000 m^3).

Strategic Petroleum Reserve (United States)

The Strategic Petroleum Reserve (SPR) is an emergency fuel store of oil maintained by the United States Department of Energy.

United States

The US SPR is the largest emergency supply in the world with the current capacity to hold up to 727 million barrels (115,600,000 m^3).

The current inventory is displayed on the SPR's website. As of May 31, 2011, the current inventory was 726.5 million barrels (115,500,000 m^3). This equates to 34 days of oil at current daily US consumption levels of 21 million barrels per day (3,300,000 m^3/d). At recent market prices ($65 a barrel as of October 2008) the SPR holds over $34.3 billion in sweet crude and approximately $51.2 billion in sour crude (assuming a $15/barrel discount for sulfur content). The total value of the crude in the SPR is approximately $85.5 billion USD. The price paid for the oil is $20.1 billion (an average of $28.42 per barrel).

Purchases of crude oil resumed in January 2009 using revenues available from the 2005 Hurricane Katrina emergency sale. The DOE purchased 10,700,000 barrels (1,700,000 m^3) at a cost of $553 million.

The United States started the petroleum reserve in 1975 after oil supplies were cut off during the 1973-74 oil embargo, to mitigate future temporary supply disruptions. According to the World Factbook, the United States imports a net 12 million barrels (1,900,000 m^3) of oil a day (MMbd), so the SPR holds about a 58-day supply. However, the maximum total withdrawal capability from the SPR is only 4.4 million barrels (700,000 m^3) per day, making it a 160 + day supply.

Facilities

The SPR management office is located in New Orleans, Louisiana.

The reserve is stored at four sites on the Gulf of Mexico, each located near a major centre of petrochemical refining and processing. Each site contains a number of artificial caverns created in salt domes below the surface.

Individual caverns within a site can be up to 1000 m below the surface, average dimensions are 60 m wide and 600 m deep, and capacity ranges from 6 to 37 million barrels (950,000 to 5,900,000 m^3). Almost $4 billion was spent on the facilities.

The decision to store in caverns was made in order to reduce costs; the Department of Energy claims it is roughly 10 times cheaper to store oil below surface with the added advantages of no leaks and a constant natural churn of the oil due to a temperature gradient in the caverns. The caverns were created by drilling down and then dissolving the salt with water.

Existing

- Bryan Mound - Freeport, Texas. 20 caverns with a storage capacity of 254 million barrels (40,400,000 m^3) with a drawdown capacity of 1.5 million barrels (240,000 m^3) per day.
- Big Hill - Winnie, Texas. Has a capacity of 160 million barrels (25,000,000 m^3) with a drawdown capacity of 1.1 million barrels (170,000 m^3) per day.

  This facility is planned to be expanded by 250 million barrels (40,000,000 m^3) with a new drawdown capacity of 1.5 million barrels (240,000 m^3) per day.
- West Hackberry - Lake Charles, Louisiana. Has a capacity of 227 million barrels (36,100,000 m^3) with a drawdown capacity of 1.3 million barrels (210,000 m^3) per day.
- Bayou Choctaw - Baton Rouge, Louisiana. Has a capacity of 76 million barrels (12,100,000 m^3) with a maximum drawdown rate of 550,000 barrels (87,000 m^3). This facility is planned to be expand to 109 million barrels (17,300,000 m^3) with a new drawdown capacity of 600,000 barrels (95,000 m^3) per day.

Future

Richton, Mississippi: This facility, if built as planned, will have a capacity of 160 million barrels (25,000,000 m^3) with a drawdown capacity of 1 million barrels (160,000 m^3) per day. The Secretary of the Energy Department, Samuel Bodman, announced the creation of this site in February 2007. This new site is currently facing some opposition.

Retired

Weeks Island - Iberia Parish, Louisiana (Decommissioned 1999) Capacity of 72 million barrels (11,400,000 m^3). This facility was a conventional room and pillar near-surface salt mine, formerly owned by Morton Salt.

In 1993, a sinkhole formed on the site, allowing fresh water to intrude into the mine. Because of the mine's construction in salt deposits, fresh water would erode the ceiling, potentially causing the structure to fail.

The mine was backfilled with salt-saturated brine. This process, which allowed for recovery of 98% of the petroleum stored in the facility, reduced the risk of further freshwater intrusion, and helped prevent the remaining oil from leaking into the aquifer that is located over the salt dome.

History

Background

Access to the reserve is determined by the conditions written into the 1975 Energy Policy and Conservation Act (EPCA), primarily to counter a severe supply interruption. The maximum removal rate, by physical constraints, is 4.4 million barrels per day (700,000 m^3/d). Oil could begin entering the marketplace 13 days after a Presidential order. The Dept. of Energy says that it has about 59 days of import protection in the SPR. This, combined with private sector inventory protection, is estimated to equal 115 days of imports.

The SPR was created following the 1973 energy crisis. The EPCA of December 22, 1975, made it policy for the U.S. to establish a reserve up to one billion barrels (159 million m^3) of petroleum. A number of existing storage sites were acquired in 1977. Construction of the first surface facilities began in June 1977. On July 21, 1977, the first oil—approximately 412,000 barrels (65,500 m^3) of Saudi Arabian light crude—was delivered to the SPR. Fill was suspended in FY 1995 to devote budget resources to refurbishing the SPR equipment and extending the life of the complex. The current SPR sites are expected to be usable until around 2025. Fill was resumed in 1999.

Filling and Suspending the SPR

On November 13, 2001, President George W. Bush announced that the SPR would be filled, saying, "The Strategic Petroleum Reserve is an important element of our Nation's energy security. To maximize long-term protection against oil supply disruptions, I am directing the Secretary of Energy to fill the SPR up to its 700 million barrel [111,000,000 m^3] capacity."The highest prior level was reached in 1994 with 592 million barrels (94,100,000 m^3). At the time of President Bush's directive, the SPR contained about 545 million barrels (86,600,000 m^3). Since the directive in 2001, the capacity of the SPR increased by 27 million barrels (4,300,000 m^3) due to natural enlargement of the salt caverns in which the reserves are stored.

The Energy Policy Act of 2005 has since directed the Secretary of Energy to fill the SPR to the full 1-billion-barrel (160,000,000 m^3) authorized capacity, a process which will require a physical expansion of the Reserve's facilities.

On August 17, 2005, the SPR reached its goal of 700 million barrels (110,000,000 m^3), or about 96% of its now-increased 727-million-barrel (115,600,000 m^3) capacity. Approximately 60% of the crude oil in the reserve is the less desirable sour (high sulfur content) variety. The oil delivered to the reserve is "royalty-in-kind" oil—royalties owed to the U.S. government by operators who acquire leases on the federally owned Outer Continental Shelf in the Gulf of Mexico.

These royalties were previously collected as cash, but in 1998 the government began testing the effectiveness of collecting royalties "in kind" - or in other words, acquiring the crude oil itself. This mechanism was adopted when refilling the SPR began, and once filling is completed, revenues from the sale of future royalties will be paid into the federal treasury.

On April 25, 2006, President Bush announced a temporary halt to petroleum deposits to the SPR as part of a four point program to alleviate high fuel prices.

On January 23, 2007, President Bush suggested in his State of the Union speech that Congress should approve expansion of the current reserve capacity to twice its current level.

In April 2008, Speaker Pelosi called on President Bush to suspend purchases of oil for the Strategic Petroleum Reserve (SPR) temporarily.

On May 12, 2008, Rep. Peter Welch (D, Vermont) and 63 co-sponsors introduced the Strategic Petroleum Reserve Fill Suspension and Consumer Protection Act bill (H.R.6022), to suspend the acquisition of petroleum for the Strategic Petroleum Reserve.

On May 16, 2008, the U.S. Department of Energy said it would halt all deliveries to the Strategic Petroleum Reserve sometime in July. This announcement came days after Congress voted to direct the Bush administration to do the same. The U.S. Department of Energy did not state when the shipments would resume.

On May 19, 2008, President Bush signed the Act passed by the Congress, which he previously opposed.

On January 2, 2009, the U.S. Energy Department said that it would begin buying approximately 12,000,000 barrels (1,900,000 m^3) of crude oil to fill the Strategic Petroleum Reserve, replenishing supplies that were sold after hurricanes Katrina and Rita in 2005. The purchase will be funded by the roughly $600 million received in 2005 from the emergency sales.

Emergency Sales to Israel

According to the 1975 Second Sinai withdrawal document signed by the United States and Israel, in an emergency the U.S. is obligated to make oil available for sale to Israel for up to 5 years.

Limitations

The Strategic Petroleum Reserve is exclusively a crude petroleum reserve, not a stockpile of refined petroleum fuels, such as gasoline, diesel and kerosene. Although there are small-scale (2,000,000 barrels) heating oil reserves in Connecticut, Rhode Island and New Jersey under the aegis of the Department of Energy (DOE), the federal government maintains no gasoline reserves on anything like the scale of the SPR.

Consequently, while the US enjoys some protection from disruptions in oil supplies, it would have to depend on other stockpiling members of the International Energy Agency for relief from any major disruption to refinery operations. Since no new refineries have been constructed in the US for thirty years, there is little excess refining capacity. This was illustrated during Hurricane Katrina, when many of the Gulf coast oil refining complexes were disrupted for some time.

There have been suggestions that the DOE should stockpile both gasoline and jet fuel, to rectify this weakness. Some countries and zones, such as Australia, have a strategic reserve of both petroleum and petroleum products. In some cases, this includes a strategic reserve of jet fuel.

The former Secretary of Energy, Samuel Bodman, has said the Department will consider refined products as part of the expansion of between 1 billion and 1.5 billion barrels (240,000,000 m^3).

SPR Drawdowns

Petroleum Sales

- 1985 - Test sale - 1.1 million barrels (170,000 m^3)
- 1990/91 - Desert Storm sale - 21 million barrels (3,300,000 m^3)
  - 4 million in August 1990 test sale
  - 17 million in January 1991 Presidentially-ordered drawdown
- 1996-97 total non-emergency sales for deficit reduction - 28 million barrels (4,500,000 m^3)
- 2005 - Hurricane Katrina sale - 11 million barrels (1,700,000 m^3) Katrina shut down 95% of crude production and 88% of natural gas output in the Gulf of Mexico. This amounted to a quarter of total U.S. output. About 735 oil and natural gas rigs and platforms had been evacuated due to the hurricane.
- 2011 - Arab Spring sale - 30 million barrels (4,800,000 m^3) to offset disruptions caused by political upheaval in Libya and

elsewhere in the Middle East. The amount was matched by IEA countries, for a total of 60 million barrels (9,500,000 m^3) released from stockpiles around the world.

Petroleum Exchanges and Loans

Note: Loans are made on a case-by-case basis to alleviate supply disruptions. Once conditions return to normal, the loan is returned to the SPR with additional oil as interest.

- April–May 1996 - 900,000 barrels (140,000 m^3) lent to ARCO to alleviate pipeline blockage.
- August 1998 - 11 million barrels (1,700,000 m^3) lent to PEMEX in return for 8.5 million barrels (1,350,000 m^3) of higher quality crude.
- June 2000 - 1 million barrels (160,000 m^3) lent to Citgo and Conoco in response to shipping channel blockage.
- July–August 2000 - 2.8 million barrels (450,000 m^3) to supply the Northeast Home Heating Oil Reserve.
- September–October 2000 - 30 million barrels (4,800,000 m^3) in response to a concern over low distillate levels in the Northeastern U.S.
- October 2002 - 296,000 barrels (47,100 m^3) lent to Shell Pipeline Company in advance of Hurricane Lili.
- September–October 2004 - 5.4 million barrels (860,000 m^3) lent to Astra Oil, ConocoPhillips, Placid Refining, Shell Oil Company, and Premcor after Hurricane Ivan.
- September–October 2005 - 9.8 million barrels (1,560,000 m^3) lent to ExxonMobil, Placid Refining, Valero, BP, Marathon Oil, and Total S.A. after Hurricane Katrina.
- January–February 2006 - 767,000 barrels (121,900 m^3) lent to Total Petrochemicals USA due to closure of the Sabine Neches ship channel to deep-draft vessels after a barge accident in the channel.
- June 2006 - 750,000 barrels (119,000 m^3) of sour crude lent to ConocoPhillips and Citgo due to the closure for several days of the Calcasieu Ship Channel caused by the release of a mixture of storm water and oil. Repaid in early October 2006.
- September 2008 - 250,000 barrels (40,000 m^3) loaned to Citgo because it could not secure crude oil in the aftermath of Hurricane Gustav.

- September 2008 - 130,000 barrels (21,000 m^3) loaned to Placid Refining's Port Allen refinery and 250,000 barrels (40,000 m^3) loaned to Marathon Oil due to disruptions from Hurricane Gustav.

Drillship

A drillship is a maritime vessel that has been fitted with drilling apparatus. It is most often used for exploratory offshore drilling of new oil or gas wells in deep water or for scientific drilling. The drillship can also be used as a platform to carry out well maintenance or completion work such as casing and tubing installation or subsea tree installations. It is often built to the design specification of the oil production company and/or investors, but can also be a modified tanker hull outfitted with a dynamic positioning system to maintain its position over the well.

The greatest advantages these modern drillships have is their ability to drill in water depths of more than 2500 meters and the time saved sailing between oilfields worldwide. Drillships are completely independent, in contrast to semi-submersibles and jackup barges.

In order to drill, a marine riser is lowered from the drillship to the seabed with a blowout preventer (BOP) at the bottom that connects to the wellhead.

Drillships are just one way to perform exploratory drilling. This function can also be performed by semi-submersibles, jackup barges, barges, or platform rigs.

The first drillship was the Cuss 1, and the fleet size has been growing ever since. By 2013 the worldwide fleet of drillships is expected to top 80 ships, more than double its size in 2009. Drillships are not only growing in size but also in capability with new technology assisting operations from academic research to ice drilling. U.S. President Barack Obama's decision in late March 2010 to expand U.S. domestic exploratory drilling seems likely to increase further developments of drillship technology.

Stena drillings 'DrillMAX ICE' being constructed by Samsung Heavy Industries in South Korea is to be the most expensive drillship built to date and is to complete exploration in Arctic regions when completed in 2012.

Chapter 3

Natural Gas and the Environment

Natural gas is an extremely important source of energy for reducing pollution and maintaining a clean and healthy environment. In addition to being a domestically abundant and secure source of energy, the use of natural gas also offers a number of environmental benefits over other sources of energy, particularly other fossil fuels. This section will discuss the environmental effects of natural gas in terms of emissions, as well as the environmental impact of the natural gas industry itself. Scroll down, or click on the links below to be transported ahead.

- Emissions from the Combustion of Natural Gas
    - o Greenhouse Gas Emissions
    - o Smog, Air Quality and Acid Rain
    - o Pollution from Industry and Electric Generation
    - o Pollution from the Transportation Sector
- The Natural Gas Industry and the Environment.

Emissions from the Combustion of Natural Gas

Natural gas is the cleanest of all the fossil fuels, as evidenced in the Environmental Protection Agency's data comparisons in the chart below, which is still current as of 2010. Composed primarily of methane, the main products of the combustion of natural gas are carbon dioxide and water vapour, the same compounds we exhale when we breathe. Coal and oil are composed of much more complex molecules, with a higher carbon ratio and higher nitrogen and sulfur contents. This means that when combusted, coal and oil release higher levels of harmful emissions, including a higher ratio of carbon emissions, nitrogen oxides (NOx), and sulfur dioxide (SO_2). Coal and fuel oil also release ash particles into the environment, substances that do not burn but instead are carried into the atmosphere and contribute to pollution. The combustion of natural gas, on the other hand, releases very small amounts of sulfur dioxide and nitrogen oxides, virtually no ash or particulate matter, and lower levels of carbon dioxide, carbon monoxide, and other reactive hydrocarbons.

Table: *Fossil Fuel Emission Levels-Pounds per Billion Btu of Energy Input*

Pollutant	***Natural Gas***	***Oil***	***Coal***
Carbon Dioxide	117,000	164,000	208,000
Carbon Monoxide	40	33	208
Nitrogen Oxides	92	448	457
Sulfur Dioxide	1	1,122	2,591
Particulates	7	84	2,744
Mercury	0.000	0.007	0.016

Source: EIA - Natural Gas Issues and Trends 1998

Natural gas, as the cleanest of the fossil fuels, can be used in many ways to help reduce the emissions of pollutants into the atmosphere. Burning natural gas in the place of other fossil fuels emits fewer harmful pollutants, and an increased reliance on natural gas can potentially reduce the emission of many of these most harmful pollutants.

Pollutants emitted in the United States, particularly from the combustion of fossil fuels, have led to the development of many pressing environmental problems. Natural gas, emitting fewer harmful chemicals into the atmosphere than other fossil fuels, can help to mitigate some of these environmental issues. These issues include:

- Greenhouse Gas Emissions
- Smog, Air Quality and Acid Rain
- Industrial and Electric Generation Emissions
- Pollution from the Transportation Sector - Natural Gas Vehicles.

Greenhouse Gas Emissions

Global warming, or the 'greenhouse effect' is an environmental issue that deals with the potential for global climate change due to increased levels of atmospheric 'greenhouse gases'. There are certain gases in our atmosphere that serve to regulate the amount of heat that is kept close to the earth's surface. Scientists theorize that an increase in these greenhouse gases will translate into increased temperatures around the globe, which would result in many disastrous environmental effects. In fact, the Intergovernmental Panel on Climate Change (IPCC) predicts in its *'Fourth Assessment Report'* released in 2007 that during the 21st century, global average temperatures are expected to rise by between 2.0 and 11.5 degrees Fahrenheit. A Fifth Assessment Report is expected to be released by the IPCC between 2013 and 2015.

The principle greenhouse gases include water vapour, carbon dioxide, methane, nitrogen oxides, and some engineered chemicals such as cholorofluorocarbons. While most of these gases occur in the atmosphere naturally, levels have been increasing due to the widespread burning of fossil fuels by growing human populations. The reduction of greenhouse gas emissions has become a primary focus of environmental programs in countries around the world.

One of the principle greenhouse gases is carbon dioxide. Although carbon dioxide does not trap heat as effectively as other greenhouse gases (making it a less potent greenhouse gas), the sheer volume of carbon dioxide emissions into the atmosphere is very high, particularly from the burning of fossil fuels. In fact, according to the Energy Information Administration in its December 2009 report *'Emissions of Greenhouse Gases'* in the United States, 81.3 percent of greenhouse gas emissions in the United States in 2008 came from energy-related carbon dioxide.

Because carbon dioxide makes up such a high proportion of U.S. greenhouse gas emissions, reducing carbon dioxide emissions can play a pivotal role in combating the greenhouse effect and global warming. The combustion of natural gas emits almost 30 percent less carbon dioxide than oil, and just under 45 percent less carbon dioxide than coal.

One issue that has arisen with respect to natural gas and the greenhouse effect is the fact that methane, the principle component of natural gas, is itself a potent greenhouse gas. Methane has an ability to trap heat almost 21 times more effectively than carbon dioxide. According to the Energy Information Administration, although methane emissions account for only 1.1 percent of total U.S. greenhouse gas emissions, they account for 8.5 percent of the greenhouse gas emissions based on global warming potential. Sources of methane emissions in the U.S. include the waste management and operations industry, the agricultural industry, as well as leaks and emissions from the oil and gas industry itself. A major study performed by the Environmental Protection Agency (EPA) and the Gas Research

Institute (GRI), now Gas Technology Institute, in 1997 sought to discover whether the reduction in carbon dioxide emissions from increased natural gas use would be offset by a possible increased level of methane emissions.

The study concluded that the reduction in emissions from increased natural gas use strongly outweighs the detrimental effects of increased methane emissions. More recently in 2011, researchers at the Carnegie Mellon University released "Life cycle greenhouse gas emissions of Marcellus shale gas", a report comparing greenhouse gas emissions from the Marcellus Shale region with emissions from coal used for electricity generation. The authors found that wells in the Marcellus region emit 20 percent to 50 percent less greenhouse gases than coal used to produce electricity.

In 1993, the natural gas industry joined with EPA in launching the Natural Gas STAR Program to reduce methane emissions. The STAR program has chronicled dramatic reductions to methane emissions, since that time:

- EPA STAR data shows a reduction in methane emissions each year for the last 16 years
- More than 904 Billion cubic feet (Bcf) of methane emissions were eliminated through the STAR program 1993-2009; and
- In 2009 alone, the program reduced methane emissions by 86 Bcf.

Thus the increased use of natural gas in the place of other, dirtier fossil fuels can serve to lessen the emission of greenhouse gases in the United States.

Smog, Air Quality and Acid Rain

Smog and poor air quality is a pressing environmental problem, particularly for large metropolitan cities. Smog, the primary constituent of which is ground level ozone, is formed by a chemical reaction of carbon monoxide, nitrogen oxides, volatile organic compounds, and heat from sunlight. As well as creating that familiar smoggy haze commonly found surrounding large cities, particularly in the summer time, smog and ground level ozone can contribute to respiratory problems ranging from temporary discomfort to long-lasting, permanent lung damage. Pollutants contributing to smog come from a variety of sources, including vehicle emissions, smokestack emissions, paints, and solvents. Because the reaction to create smog requires heat, smog problems are the worst in the summertime.

The use of natural gas does not contribute significantly to smog formation, as it emits low levels of nitrogen oxides, and virtually no particulate matter. For this reason, it can be used to help combat smog formation in those areas where ground level air quality is poor. The main sources of nitrogen oxides are electric utilities, motor vehicles, and industrial plants.

Increased natural gas use in the electric generation sector, a shift to cleaner natural gas vehicles, or increased industrial natural gas use, could all serve to combat smog production, especially in urban centers where it is needed the most. Particularly in the summertime, when natural gas demand is lowest and smog problems are the greatest, industrial plants and electric generators could use natural gas to fuel their operations instead of other, more polluting fossil fuels. This would effectively reduce the emissions of smog causing chemicals, and result in clearer, healthier air around urban centers.

Particulate emissions also cause the degradation of air quality in the United States. These particulates can include soot, ash, metals, and other airborne particles. A study by the Union of Concerned Scientists in 1998, entitled 'Cars and Trucks and Air Pollution', showed that the risk of premature death for residents in areas with high airborne particulate matter was 26 percent greater than for those in areas with low particulate levels.

Natural gas emits virtually no particulates into the atmosphere: in fact, emissions of particulates from natural gas combustion are 90 percent lower than from the combustion of oil, and 99 percent lower than burning coal. Thus increased natural gas use in place of other dirtier hydrocarbons can help to reduce particulate emissions in the U.S. Current consequences stemming from global warming raised by the Union of Concerned Scientists can be found on their site.

Acid rain is another environmental problem that affects much of the Eastern United States, damaging crops, forests, wildlife populations, and causing respiratory and other illnesses in humans. Acid rain is formed when sulfur dioxide and nitrogen oxides react with water vapour and other chemicals in the presence of sunlight to form various acidic compounds in the air. The principle source of acid rain-causing pollutants, sulfur dioxide and nitrogen oxides, are coal fired power plants. Since natural gas emits virtually no sulfur dioxide, and up to 80 percent less nitrogen oxides than the combustion of coal, increased use of natural gas could provide for fewer acid rain causing emissions.

Industrial and Electric Generation Emissions

Pollutant emissions from the industrial sector and electric utilities contribute greatly to environmental problems in the United States. The use of natural gas to power both industrial boilers and processes and the generation of electricity can significantly improve the emissions profiles for these two sectors.

Natural gas is becoming an increasingly important fuel in the generation of electricity. As well as providing an efficient, competitively priced fuel for the generation of electricity, the increased use of natural gas allows for the improvement in the emissions profile of the electric generation industry. According to the National Environmental Trust (NET), now Pew Charitable Trusts (PEW), in their 2002 publication entitled 'Cleaning up Air Pollution from America's Power Plants', power plants in the U.S. account for 67 percent of sulfur dioxide emissions, 40 percent of carbon dioxide emissions, 25 percent of nitrogen oxide emissions, and 34 percent of mercury emissions. Coal fired power plants are the greatest contributors to these types of emissions. In fact, according to World Watch Report 184, natural gas combined cycle power plants emit half as much carbon dioxide as modern super critical coal plants.

Natural gas-fired electric generation and natural gas-powered industrial applications offer a variety of environmental benefits and environmentally friendly uses, including:

- Fewer Emissions - Combustion of natural gas, used in the generation of electricity, industrial boilers, and other applications, emits lower levels of NOx, CO_2, and particulate emissions, and virtually no SO_2 and mercury emissions. Natural gas can be used in place of, or in addition to, other fossil fuels, including coal, oil, or petroleum coke, which emit significantly higher levels of these pollutants.
- Reduced Sludge – Coal-fired power plants and industrial boilers that use scrubbers to reduce SO_2 emissions levels generate thousands of tons of harmful sludge. Combustion of natural gas emits extremely low levels of SO_2, eliminating the need for scrubbers, and reducing the amounts of sludge associated with power plants and industrial processes.
- Reburning - This process involves injecting natural gas into coal or oil fired boilers. The addition of natural gas to the fuel mix can result in NOx emission reductions of 50 to 70 percent, and SO_2 emission reductions of 20 to 25 percent.

- Cogeneration - The production and use of both heat and electricity can increase the energy efficiency of electric generation systems and industrial boilers, which translates to the combustion of less fuel and the emission of fewer pollutants. Natural gas is the preferred choice for new cogeneration applications.
- Combined Cycle Generation – Combined-cycle generation units generate electricity and capture normally wasted heat energy, using it to generate more electricity. Like cogeneration applications, this increases energy efficiency, uses less fuel, and thus produces fewer emissions. Natural gas-fired combined-cycle generation units can be up to 60 percent energy efficient, whereas coal and oil generation units are typically only 30 to 35 percent efficient.
- Fuel Cells - Natural gas fuel cell technologies are in development for the generation of electricity. Fuel cells are sophisticated devices that use hydrogen to generate electricity, much like a battery. No emissions are involved in the generation of electricity from fuel cells, and natural gas, being a hydrogen rich source of fuel, can be used. Although still under development, widespread use of fuel cells could in the future significantly reduce the emissions associated with the generation of electricity.

Essentially, electric generation and industrial applications that require energy, particularly for heating, use the combustion of fossil fuels for that energy. Because of its clean burning nature, the use of natural gas wherever possible, either in conjunction with other fossil fuels, or instead of them, can help to reduce the emission of harmful pollutants.

According to the Congressional Research Service's 2010 report: *"Displacing Coal with Generation from Existing Natural-Gas Fired Power Plants,"* if natural-gas combined cycle plants utilization were to be doubled from 42 percent capacity factor to 85 percent, then the amount of power generated would displace 19 percent of the CO2 emissions attributed to coal-fired electricity generation.

The transportation sector (particularly cars, trucks, and buses) is one of the greatest contributors to air pollution in the United States. Emissions from vehicles contribute to smog, low visibility, and various greenhouse gas emissions. According to the Department of Energy (DOE), about half of all air pollution and more than 80 percent of air pollution in cities are produced by cars and trucks in the United States. Currently, automobile manufacturers are under pressure to produce more environmentally friendly vehicles.

Natural gas can be used in the transportation sector to cut down on these high levels of pollution from gasoline and diesel powered cars, trucks, and buses. According to the EPA, compared to traditional vehicles, vehicles operating on compressed natural gas have reductions in carbon monoxide emissions of 90 to 97 percent, and reductions in carbon dioxide emissions of 25 percent. Nitrogen oxide emissions can be reduced by 35 to 60 percent, and other non-methane hydrocarbon emissions could be reduced by as much as 50 to 75 percent. In addition, because of the relatively simple makeup of natural gas in comparison to traditional vehicle fuels, there are fewer toxic and carcinogenic emissions from natural gas vehicles, and virtually no particulate emissions. Thus the environmentally friendly attributes of natural gas may be used in the transportation sector to reduce air pollution. Natural gas vehicles represent a growing segment of the transportation sector. According to the Natural Gas Vehicle Coalition, the use of natural gas for vehicles doubled between 2003 and 2009. Over 100,000 natural gas vehicles are currently on US roads. A large portion of those vehicles are transit buses, which account for nearly 62 percent of all natural gas vehicles.

Natural gas is the cleanest of the fossil fuels, and thus its many applications can serve to decrease harmful pollution levels from all sectors, particularly when used together with or replacing other fossil fuels. The natural gas industry itself is also committed to ensuring that the process of producing natural gas is as environmentally-friendly as possible. The Natural Gas Vehicle Coalition has more information regarding natural gas-powered vehicles.

Chapter 4

The Transportation of Natural Gas

The efficient and effective movement of natural gas from producing regions to consumption regions requires an extensive and elaborate transportation system. In many instances, natural gas produced from a particular well will have to travel a great distance to reach its point of use. The transportation system for natural gas consists of a complex network of pipelines, designed to quickly and efficiently transport natural gas from its origin, to areas of high natural gas demand. Transportation of natural gas is closely linked to its storage: should the natural gas being transported not be immediately required, it can be put into storage facilities for when it is needed. There are three major types of pipelines along the transportation route: the gathering system, the interstate pipeline system, and the distribution system. The gathering system consists of low pressure, small diameter pipelines that transport raw natural gas from the wellhead to the processing plant. Should natural gas from a particular well have high sulfur and carbon dioxide contents (sour gas), a specialized sour gas gathering pipe must be installed. Sour gas is corrosive, thus its transportation from the wellhead to the sweetening plant must be done carefully. Review the treatment and processing of natural gas.

Pipelines can be characterized as interstate or intrastate. Interstate pipelines are similar to in the interstate highway system: they carry natural gas across state boundaries, in some cases clear across the country. Intrastate pipelines, on the other hand, transport natural gas within a particular state. This section will cover only the fundamentals of interstate natural gas pipelines, however the technical and operational details discussed are essentially the same for intrastate pipelines.

Interstate Natural Gas Pipelines

The interstate natural gas pipeline network transports processed natural gas from processing plants in producing regions to those areas with high natural gas requirements, particularly large, populated urban areas. As can be seen, the pipeline network extends across the entire country.

Interstate pipelines are the 'highways' of natural gas transmission. Natural gas that is transported through interstate pipelines travels at high pressure in the pipeline, at pressures anywhere from 200 to 1500 pounds per square inch (psi). This reduces the volume of the natural gas being transported (by up to 600 times), as well as propelling natural gas through the pipeline.

This section will cover the components of the interstate pipeline system, the construction of pipelines, and pipeline inspection and safety.

Pipeline Components

Interstate pipelines consist of a number of components that ensure the efficiency and reliability of a system that delivers such an important energy source year-round, twenty four hours a day, and includes a number of different components.

Transmission Pipes

Transmission pipes can measure anywhere from 6 to 48 inches in diameter, depending on their function. Certain component pipe sections can even consist of small diameter pipe, as small as 0.5 inches in diameter. However, this small diameter pipe is usually used only in gathering and distribution systems. Mainline transmission pipes, the principle pipeline in a given system, are usually between 16 and 48 inches in diameter. Lateral pipelines, which deliver natural gas to or from the mainline, are typically between 6 and 16 inches in diameter. Most major interstate pipelines are between 24 and 36 inches in diameter. The actual pipeline itself, commonly called 'line pipe', consists of a strong carbon steel material, engineered to meet standards set by the American Petroleum Institute (API). In contrast, some distribution pipe is made of highly advanced plastic, because of the need for flexibility, versatility and the ease of replacement.

Transmission pipelines are produced in steel mills, which are sometimes specialized to produce only pipeline. There are two different production techniques, one for small diameter pipes and one for large diameter pipes. For large diameter pipes, from 20 to 42 inches in diameter, the pipes are produced from sheets of metal which are folded into a tube shape, with the ends welded together to form a pipe section. Small diameter pipe, on the other hand, can be produced seamlessly. This involves heating a metal bar to very high temperatures, then punching a hole through the middle of the bar to produce a hollow tube. In either case, the pipe is tested before being shipped from the steel mill, to ensure that it can meet the pressure and strength standards for transporting natural gas.

Line pipe is also covered with a specialized coating to ensure that it does not corrode once placed in the ground. The purpose of the coating is to protect the pipe from moisture, which causes corrosion and rusting. There are a number of different coating techniques. In the past, pipelines were coated with specialized coal tar enamel. Today, pipes are often protected with what is known as a fusion bond epoxy, which gives the pipe a noticeable light blue colour. In addition, cathodic protection is often used; which is a technique of running an electric current through the pipe to ward off corrosion and rusting.

Compressor Stations

As mentioned, natural gas is highly pressurized as it travels through an interstate pipeline. To ensure that the natural gas flowing through any one pipeline remains pressurized, compression of this natural gas is required periodically along the pipe. This is accomplished by compressor stations, usually placed at 40 to 100 mile intervals along the pipeline. The natural gas enters the compressor station, where it is compressed by either a turbine, motor, or engine.

Turbine compressors gain their energy by using up a small proportion of the natural gas that they compress. The turbine itself serves to operate a centrifugal compressor, which contains a type of fan that compresses and pumps the natural gas through the pipeline. Some compressor stations are operated by using an electric motor to turn the same type of centrifugal compressor. This type of compression does not require the use of any of the natural gas from the pipe, however it does require a reliable source of electricity nearby. Reciprocating natural gas engines are also used to power some compressor stations. These engines resemble a very large automobile engine, and are powered by natural gas from the pipeline. The combustion of the natural gas powers pistons on the outside of the engine, which serves to compress the natural gas.

In addition to compressing natural gas, compressor stations also usually contain some type of liquid separator, much like the ones used to dehydrate natural gas during its processing. Usually, these separators consist of scrubbers and filters that capture any liquids or other unwanted particles from the natural gas in the pipeline. Although natural gas in pipelines is considered 'dry' gas, it is not uncommon for a certain amount of water and hydrocarbons to condense out of the gas stream while in transit. The liquid separators at compressor stations ensure that the natural gas in the pipeline is as pure as possible, and usually filter the gas prior to compression.

Metering Stations

In addition to compressing natural gas to reduce its volume and push it through the pipe, metering stations are placed periodically along interstate natural gas pipelines. These stations allow pipeline companies to monitor the natural gas in their pipes. Essentially, these metering stations measure the flow of gas along the pipeline, and allow pipeline companies to 'track' natural gas as it flows along the pipeline. These metering stations employ specialized meters to measure the natural gas as it flows through the pipeline, without impeding its movement.

Valves

Interstate pipelines include a great number of valves along their entire length. These valves work like gateways; they are usually open and allow natural gas to flow freely, or they can be used to stop gas flow along a certain section of pipe. There are many reasons why a pipeline may need to restrict gas flow in certain areas. For example, if a section of pipe requires replacement or maintenance, valves on either end of that section of pipe can be closed to allow engineers and work crews safe access. These large valves can be placed every 5 to 20 miles along the pipeline, and are subject to regulation by safety codes.

Control Stations and SCADA Systems

Natural gas pipeline companies have customers on both ends of the pipeline - the producers and processors that input gas into the pipeline, and the consumers and local gas utilities that take gas out of the pipeline. In order to manage the natural gas that enters the pipeline, and to ensure that all customers receive timely delivery of their portion of this gas, sophisticated control systems are required to monitor the gas as it travels through all sections of what could be a very lengthy pipeline network. To accomplish this task of monitoring and controlling the natural gas that is travelling through the pipeline, centralized gas control stations collect, assimilate, and manage data received from monitoring and compressor stations all along the pipe.

Most of the data that is received by a control station is provided by Supervisory Control and Data Acquisition (SCADA) systems. These systems are essentially sophisticated communications systems that take measurements and collect data along the pipeline (usually in a metering or compressor stations and valves) and transmit it to the centralized control station. Flow rate through the pipeline, operational status, pressure, and temperature readings may all be used to assess the status of the pipeline at any one time. These systems also work in real time, meaning that there is little lag time between the measurements taken along the pipeline and their transmission to the control station.

The data is relayed to a centralized control station, allowing pipeline engineers to know exactly what is happening along the pipeline at all times. This enables quick reactions to equipment malfunctions, leaks, or any other unusual activity along the pipeline. Some SCADA systems also incorporate the ability to remotely operate certain equipment along the pipeline, including compressor stations, allowing engineers in a centralized control centre to immediately and easily adjust flow rates in the pipeline.

Pipeline Construction

As natural gas use increases, so does the need to have transportation infrastructure in place to supply the increased demand. This means that pipeline companies are constantly assessing the flow of natural gas across the U.S., and building pipelines to allow transportation of natural gas to those areas that are underserved.

Constructing natural gas pipelines requires a great deal of planning and preparation. In addition to actually building the pipeline, several permitting and regulatory processes must be completed. In many cases, prior to beginning the permitting and land access processes, natural gas pipeline companies prepare a feasibility analysis to ensure that an acceptable route for the pipeline exists that provides the least impact to the environment and public infrastructure already in place.

Assuming a pipeline company obtains all the required permits and satisfies all of the regulatory requirements, construction of the pipe may begin. Extensive surveying of the intended route is completed, both aerial and land based, to ensure that no surprises pop up during actual assembly of the pipeline.

Installing a pipeline is much like an assembly line process, with sections of the pipeline being completed in stages. First, the path of the pipeline is cleared of all removable impediments, including trees, boulders, brush, and anything else that may prohibit the construction.

Once the pipeline's path has been cleared sufficiently to allow construction equipment to gain access, sections of pipes are laid out along the intended path, a process called 'stringing' the pipe. These pipe sections are commonly from 40 to 80 feet long, and are specific to their destination. That is, certain areas have different requirements for coating material and pipe thickness.

Once the pipe is in place, trenches are dug alongside the laid out pipe. These trenches are typically five to six feet deep, as the regulations require the pipe to be at least 30 inches below the surface. In certain areas, however, including road crossings and bodies of water, the pipe is buried even deeper. Once the trenches are dug, the pipe is assembled and contoured.

This includes welding the sections of pipe together into one continuous pipeline, and bending it slightly, if needed, to fit the contour of the pipeline's path. Coating is applied to the ends of the pipes. The coating applied at a coating mill typically leaves the ends of the pipe clean, so as not to interfere with welding. Finally, the entire coating of the pipe is inspected to ensure that it is free from defects.

Once the pipe is welded, bent, coated, and inspected it can be lowered into the previously-dug trenches. This is done with specialized construction equipment acting to lift the pipe in a level manner and lower it into the trench. Once lowered into the ground, the trench is filled in carefully, to ensure that the pipe and its coating retain their integrity.

The last step in pipeline construction is the hydrostatic test. This consists of running water, at pressures higher than will be needed for natural gas transportation, through the entire length of the pipe. This serves as a test to ensure that the pipeline is strong enough, and absent of any leaks of fissures, before natural gas is pumped through the pipeline.

Laying pipe across streams or rivers can be accomplished in one of two ways. Open cut crossing involves the digging of trenches on the floor of the river to house the pipe. When this is done, the pipe itself is usually fitted with a concrete casing, which both ensures that the pipe stays on the bottom of the river and adds an extra protective coating to prevent any natural gas leaks into the water. Alternatively, a form of directional drilling may be employed, in which a 'tunnel' is drilled under the river through which the pipe may be passed. The same techniques are used for road crossings - either an open trench is excavated across the road and replaced once the pipe is installed, or a tunnel may be drilled underneath the road.

Once the pipeline has been installed and covered, extensive efforts are taken to restore the pipeline's pathway to its original state, or to mitigate any environmental or other impacts that may have occurred during the construction process.

These steps often include replacing topsoil, fences, irrigation canals, and anything else that may have been removed or upset during the construction process.

Pipeline Inspection and Safety

In order to ensure the efficient and safe operation of the extensive network of natural gas pipelines, pipeline companies routinely inspect their pipelines for corrosion and defects. This is done through the use of sophisticated pieces of equipment known as 'smart pigs.' Smart pigs are intelligent robotic devices that are propelled down pipelines to evaluate the interior of the pipe. Smart pigs can test pipe thickness, and roundness, check for signs of corrosion, detect minute leaks, and any other defect along the interior of the pipeline that may either impede the flow of gas, or pose a potential safety risk to the operation of the pipeline. Sending a smart pig down a pipeline is fittingly known as 'pigging' the pipeline.

In addition to inspection with smart pigs, there are a number of safety precautions and procedures in place to minimize the risk of accidents. In fact, the transportation of natural gas is one of the safest ways of transporting energy, mostly due to the fact that the infrastructure is fixed, and buried underground. According to the Department of Transportation (DOT), pipelines are the safest method of transporting petroleum and natural gas.

While there are in excess of 100 deaths per year associated with electric transmission lines, according to the DOT's Office of Pipeline Safety in 2009, there were 0 deaths associated with transmission pipelines, and 10 deaths associated with distribution systems. To learn more about pipeline safety, visit the DOT's Office of Pipeline Safety.

A few of the safety precautions associated with natural gas pipelines include:

- *Aerial Patrols* - Planes are used to ensure no construction activities are taking place too close to the route of the pipeline, particularly in residential areas. Unauthorized construction and digging is the primary threat to pipeline safety, according to INGAA
- *Leak Detection* - Natural gas detecting equipment is periodically used by pipeline personnel on the surface to check for leaks.

This is especially important in areas where the natural gas is not odorized.

- *Pipeline Markers* - Signs on the surface above natural gas pipelines indicate the presence of underground pipelines to the public, to reduce the chance of any interference with the pipeline.
- *Gas Sampling* - Routine sampling of the natural gas in pipelines ensures its quality, and may also indicate corrosion of the interior of the pipeline, or the influx of contaminants.
- *Preventative Maintenance* - This involves the testing of valves and the removal of surface impediments to pipeline inspection.
- *Emergency Response* - Pipeline companies have extensive emergency response teams that train for the possibility of a wide range of potential accidents and emergencies.
- *The One Call Program* - All 50 states have instituted what is known as a 'one call' program, which provides excavators, construction crews, and anyone interested in digging into the ground around a pipeline with a single phone number that may be called when any excavation activity is planned. This call alerts the pipeline company, which may flag the area, or even send representatives to monitor the digging. The national 3-digit number for one call is "811."

While large interstate natural gas pipelines transport natural gas from the processing regions to the consuming regions and may serve large wholesale users such as industrial or power generation customers directly, it is the distribution system that actually delivers natural gas to most retail customers, including residential natural gas users.

Storage of Natural Gas

Natural gas, like most other commodities, can be stored for an indefinite period of time. The exploration, production, and transportation of natural gas takes time, and the natural gas that reaches its destination is not always needed right away, so it is injected into underground storage facilities. These storage facilities can be located near market centers that do not have a ready supply of locally produced natural gas.

Traditionally, natural gas has been a seasonal fuel. That is, demand for natural gas is usually higher during the winter, partly because it is used for heat in residential and commercial settings. Stored natural gas plays a vital role in ensuring that any excess supply delivered during the summer months is available to meet the increased demand of the winter months. However, with the recent trend towards natural gas fired electric generation, demand for natural gas during the summer months is now increasing (due to the demand for electricity to power air conditioners and the like). Natural gas in storage also serves as insurance against any unforeseen accidents, natural disasters, or other occurrences that may affect the production or delivery of natural gas.

Natural gas storage plays a vital role in maintaining the reliability of supply needed to meet the demands of consumers. Historically, when natural gas was a regulated commodity, storage was part of the bundled product sold by the pipelines to distribution utilities. This all changed in 1992 with the introduction of the Federal Energy Regulatory Commission's (FERC) Order 636, which opened up the natural gas market to deregulation.

Essentially, this meant that where natural gas storage was required prior to Order 636 for the operational requirements of the pipelines in meeting the needs of the utilities, it is now available to anyone seeking storage for commercial purposes or operational requirements. Storage used to serve only as a buffer between transportation and distribution, to ensure adequate supplies of natural gas were in place for seasonal demand shifts, and unexpected demand surges. Now, in addition to serving those purposes, natural gas storage is also used by industry participants for commercial reasons; storing gas when prices are low, and withdrawing and selling it when prices are high, for instance.

According to the Energy Information Administration (EIA), as of 2000 there was 3.899 Trillion cubic feet (Tcf) of working gas storage capacity in the United States. To learn more about the current uses of storage, including statistics and forecasts relating to natural gas storage.

Base Load vs. Peak Load Storage

There are basically two uses for natural gas in storage facilities: meeting base load requirements, and meeting peak load requirements. As mentioned, natural gas storage is required for two reasons: meeting seasonal demand requirements, and as insurance against unforeseen supply disruptions. Base load storage capacity is used to meet seasonal demand increases. Base load facilities are capable of holding enough natural gas to satisfy long term seasonal demand requirements. Typically, the turn-over rate for natural gas in these facilities is a year; natural gas is generally injected during the summer (non-heating season), which usually runs from April through October, and withdrawn during the winter (heating season), usually from November to March. These reservoirs are larger, but their delivery rates are relatively low, meaning the natural gas that can be extracted each day is limited. Instead, these facilities provide a prolonged, steady supply of natural gas. Depleted gas reservoirs are the most common type of base load storage facility.

Peak load storage facilities, on the other hand, are designed to have high-deliverability for short periods of time, meaning natural gas can be withdrawn from storage quickly should the need arise. Peak load facilities are intended to meet sudden, short-term demand increases. These facilities cannot hold as much natural gas as base load facilities; however, they can deliver smaller amounts of gas more quickly, and can also be replenished in a shorter amount of time than base load facilities. While base load facilities have long term injection and withdrawal seasons, turning over the natural gas in the facility about once per year, peak load facilities can have turn over rates as short as a few days or weeks. Salt caverns are the most common type of peak load storage facility, although aquifers may be used to meet these demands as well.

Natural gas is usually stored underground, in large storage reservoirs. There are three main types of underground storage: depleted gas reservoirs, aquifers, and salt caverns. In addition to underground storage, however, natural gas can be stored as liquefied natural gas (LNG). LNG allows natural gas to be shipped and stored in liquid form, meaning it takes up much less space than gaseous natural gas.

Types of Underground Storage

Underground natural gas storage fields grew in popularity shortly after World War II. At the time, the natural gas industry noted that seasonal demand increases could not feasibly be met by pipeline delivery alone. In order to meet seasonal demand increases, the deliverability of pipelines (and thus their size), would have to increase dramatically. However, the technology required to construct such large pipelines to consuming regions was, at the time, unattainable and unfeasible. In order to be able to meet seasonal demand increases, underground storage fields were the only option.

As mentioned, there are three main types of underground natural gas storage facilities. Specific characteristics of depleted reservoirs, aquifers, and salt caverns may be found below. Essentially, any underground storage facility is reconditioned before injection, to create a sort of storage vessel underground. Natural gas is injected into the formation, building up pressure as more natural gas is added. In this sense, the underground formation becomes a sort of pressurized natural gas container.

As with newly drilled wells, the higher the pressure in the storage facility, the more readily gas may be extracted. Once the pressure drops to below that of the wellhead, there is no pressure differential left to push the natural gas out of the storage facility. This means that, in any underground storage facility, there is a certain amount of gas that may never be extracted. This is known as physically unrecoverable gas; it is permanently embedded in the formation.

In addition to this physically unrecoverable gas, underground storage facilities contain what is known as 'base gas' or 'cushion gas'. This is the volume of gas that must remain in the storage facility to provide the required pressurization to extract the remaining gas. In the normal operation of the storage facility, this cushion gas remains underground; however a portion of it may be extracted using specialized compression equipment at the wellhead.

'Working gas' is the volume of natural gas in the storage reservoir that can be extracted during the normal operation of the storage facility. This is the natural gas that is being stored and withdrawn; the capacity of storage facilities normally refers to their working gas capacity. At the beginning of a withdrawal cycle, the pressure inside the storage facility is at its highest; meaning working gas can be withdrawn at a high rate. As the volume of gas inside the storage facility drops, pressure (and thus deliverability) in the storage facility also decreases. Periodically, underground storage facility operators may reclassify portions of working gas as base gas after evaluating the operation of their facilities.

The graphs shown indicate the working gas capacity and deliverability of natural gas storage facilities in the U.S. as of 2001. It can be seen that depleted reservoirs constitute the majority of working gas capacity and deliverability. However, the high deliverability of salt caverns is shown by the high daily deliverability relative to working gas capacity.

Depleted Gas Reservoirs

The first instance of natural gas successfully being stored underground occurred in Weland County, Ontario, Canada, in 1915. This storage facility used a depleted natural gas well that had been reconditioned into a storage field. In the United States, the first storage facility was developed just south of Buffalo, New York. By 1930, there were nine storage facilities in six different states. Prior to 1950, virtually all natural gas storage facilities were in depleted reservoirs.

The most prominent and common form of underground storage consists of depleted gas reservoirs. Depleted reservoirs are those formations that have already been tapped of all their recoverable natural gas.

This leaves an underground formation, geologically capable of holding natural gas. In addition, using an already developed reservoir for storage purposes allows the use of the extraction and distribution equipment left over from when the field was productive. Having this extraction network in place reduces the cost of converting a depleted reservoir into a storage facility. Depleted reservoirs are also attractive because their geological characteristics are already well known. Of the three types of underground storage, depleted reservoirs, on average, are the cheapest and easiest to develop, operate, and maintain.

The factors that determine whether or not a depleted reservoir will make a suitable storage facility are both geographic and geologic. Geographically, depleted reservoirs must be relatively close to consuming regions.

They must also be close to transportation infrastructure, including trunk pipelines and distribution systems. While depleted reservoirs are numerous in the U.S., they are more abundantly available in producing regions. In regions without depleted reservoirs, like the upper Midwest, one of the other two storage options is required.

Geologically, depleted reservoir formations must have high permeability and porosity. The porosity of the formation determines the amount of natural gas that it may hold, while its permeability determines the rate at which natural gas flows through the formation, which in turn determines the rate of injection and withdrawal of working gas. In certain instances, the formation may be stimulated to increase permeability.

In order to maintain pressure in depleted reservoirs, about 50 percent of the natural gas in the formation must be kept as cushion gas. However, depleted reservoirs, having already been filled with natural gas and hydrocarbons, do not require the injection of what will become physically unrecoverable gas; that gas already exists in the formation.

Aquifers

Aquifers are underground porous, permeable rock formations that act as natural water reservoirs. However, in certain situations, these water containing formations may be reconditioned and used as natural gas storage facilities.

As they are more expensive to develop than depleted reservoirs, these types of storage facilities are usually used only in areas where there are no nearby depleted reservoirs. Traditionally, these facilities are operated with a single winter withdrawal period, although they may be used to meet peak load requirements as well.

Aquifers are the least desirable and most expensive type of natural gas storage facility for a number of reasons. First, the geological characteristics of aquifer formations are not as thoroughly known, as with depleted reservoirs. A significant amount of time and money goes into discovering the geological characteristics of an aquifer, and determining its suitability as a natural gas storage facility. Seismic testing must be performed, much like is done for the exploration of potential natural gas formations.

The area of the formation, the composition and porosity of the formation itself, and the existing formation pressure must all be discovered prior to development of the formation. In addition, the capacity of the reservoir is unknown, and may only be determined once the formation is further developed.

In order to develop a natural aquifer into an effective natural gas storage facility, all of the associated infrastructure must also be developed. This includes installation of wells, extraction equipment, pipelines, dehydration facilities, and possibly compression equipment. Since aquifers are naturally full of water, in some instances powerful injection equipment must be used, to allow sufficient injection pressure to push down the resident water and replace it with natural gas.

While natural gas being stored in aquifers has already undergone all of its processing, upon extraction from a water bearing aquifer formation the gas typically requires further dehydration prior to transportation, which requires specialized equipment near the wellhead.

Aquifer formations do not have the same natural gas retention capabilities as depleted reservoirs. This means that some of the natural gas that is injected escapes from the formation, and must be gathered and extracted by 'collector' wells, specifically designed to pick up gas that may escape from the primary aquifer formation.

In addition to these considerations, aquifer formations typically require a great deal more 'cushion gas' than do depleted reservoirs. Since there is no naturally occurring gas in the formation to begin with, a certain amount of natural gas that is injected will ultimately prove physically unrecoverable.

In aquifer formations, cushion gas requirements can be as high as 80 percent of the total gas volume. While it is possible to extract cushion gas from depleted reservoirs, doing so from aquifer formations could have negative effects, including formation damage. As such, most of the cushion gas that is injected into any one aquifer formation may remain unrecoverable, even after the storage facility is shut down. Most aquifer storage facilities were developed when the price of natural gas was low, meaning this cushion gas was not very expensive to give up. However, with higher prices, aquifer formations are increasingly expensive to develop.

All of these factors mean that developing an aquifer formation as a storage facility can be time consuming and expensive. In some instances, aquifer development can take 4 years, which is more than twice the time it takes to develop depleted reservoirs as storage facilities. In addition to the increased time and cost of aquifer storage, there are also environmental restrictions to using aquifers as natural gas storage. In the early 1980's the Environmental Protection Agency (EPA) set certain rules and restrictions on the use of aquifers as natural gas storage facilities. These restrictions are intended to reduce the possibility of fresh water contamination.

Salt Caverns

Underground salt formations offer another option for natural gas storage. These formations are well suited to natural gas storage in that salt caverns, once formed, allow little injected natural gas to escape from the formation unless specifically extracted. The walls of a salt cavern also have the structural strength of steel, which makes it very resilient against reservoir degradation over the life of the storage facility.

Essentially, salt caverns are formed out of existing salt deposits. These underground salt deposits may exist in two possible forms: salt domes, and salt beds. Salt domes are thick formations created from natural salt deposits that, over time, leach up through overlying sedimentary layers to form large dome-type structures.

They can be as large as a mile in diameter, and 30,000 feet in height. Typically, salt domes used for natural gas storage are between

6,000 and 1,500 feet beneath the surface, although in certain circumstances they can come much closer to the surface. Salt beds are shallower, thinner formations.

These formations are usually no more than 1,000 feet in height. Because salt beds are wide, thin formations, once a salt cavern is introduced, they are more prone to deterioration, and may also be more expensive to develop than salt domes.

Once a suitable salt dome or salt bed deposit is discovered, and deemed suitable for natural gas storage, it is necessary to develop a 'salt cavern' within the formation. Essentially, this consists of using water to dissolve and extract a certain amount of salt from the deposit, leaving a large empty space in the formation. This is done by drilling a well down into the formation, and cycling large amounts of water through the completed well. This water will dissolve some of the salt in the deposit, and be cycled back up the well, leaving a large empty space that the salt used to occupy. This process is known as 'salt cavern leaching'.

Salt cavern leaching is used to create caverns in both types of salt deposits, and can be quite expensive. However, once created, a salt cavern offers an underground natural gas storage vessel with very high deliverability. In addition, cushion gas requirements are the lowest of all three storage types, with salt caverns only requiring about 33 percent of total gas capacity to be used as cushion gas.

Salt cavern storage facilities are primarily located along the Gulf Coast, as well as in the northern states, and are best suited for peak load storage. Salt caverns are typically much smaller than depleted gas reservoirs and aquifers, in fact underground salt caverns usually take up only one one-hundredth of the acreage taken up by a depleted gas reservoir.

As such, salt caverns cannot hold the volume of gas necessary to meet base load storage requirements. However, deliverability from salt caverns is typically much higher than for either aquifers or depleted reservoirs. Therefore natural gas stored in a salt cavern may be more readily (and quickly) withdrawn, and caverns may be replenished with natural gas more quickly than in either of the other types of storage facilities. Moreover, salt caverns can readily begin flowing gas on as little as one hour's notice, which is useful in emergency situations or during unexpected short term demand surges. Salt caverns may also be replenished more quickly than other types of underground storage facilities.

Location of Natural Gas Storage Facilities

Storage facilities are most concentrated in the consuming north east region of the country, but can be found nationwide.

Natural Gas Distribution

Distribution is the final step in delivering natural gas to customers. While some large industrial, commercial, and electric generation customers receive natural gas directly from high capacity interstate and intrastate pipelines (usually contracted through natural gas marketing companies), most other users receive natural gas from their local gas utility, also called a local distribution company (LDC). LDCs are regulated utilities involved in the delivery of natural gas to consumers within a specific geographic area. There are two basic types of natural gas utilities: those owned by investors, and public gas systems owned by local governments.

Local distribution companies typically transport natural gas from delivery points located on interstate and intrastate pipelines to households and businesses through thousands of miles of small-diameter distribution pipe. The delivery point where the natural gas is transferred from a transmission pipeline to the local gas utility is often termed the 'citygate', and is an important market centre for the pricing of natural gas in large urban areas. Typically, Utilities take ownership of the natural gas at the citygate, and deliver it to each individual customer's meter.

This requires an extensive network of small-diameter distribution pipe. The U.S. Department of Transportation's Pipeline and Hazardous Materials Safety Administration reports that there are just over 2 million miles of distribution pipe in the U.S., including city mains and service pipelines that connect each meter to the main.

Because of the transportation infrastructure required to move natural gas to many diverse customers across a reasonably wide geographic area, distribution costs typically make up about half of natural gas costs for households and small volume customers. While large pipelines can reduce unit costs by transmitting large volumes of natural gas, distribution companies must deliver relatively small volumes to many more different locations. According to the Energy Information Administration (EIA), transmission and distribution costs represented about half of a typical residential natural gas customer's monthly gas utility bill in 2009, with costs of the physical natural gas commodity itself representing the other half.

Delivery of Natural Gas

The delivery of natural gas to its point of end use by a distribution utility is much like the transportation of natural gas discussed in the transportation section. However, distribution involves moving smaller volumes of gas at much lower pressures over shorter distances to a great number of individual users. Smaller-diameter pipe also is used to transport natural gas from the citygate to individual consumers.

The natural gas is periodically compressed to ensure pipeline flow, although local compressor stations are typically smaller than those used for interstate transportation.

Because of the smaller volumes of natural gas to be moved, as well as the small-diameter pipe that is used, the pressure required to move natural gas through the distribution network is much lower than that found in the transmission pipelines. While natural gas travelling through interstate pipelines may be compressed to as much as 1,500 pounds per square inch (psi), natural gas travelling through the distribution network requires as little as 3 psi of pressurization and is as low as ¼ psi at the customer's meter.

The natural gas to be distributed is typically depressurized at or near the citygate, as well as scrubbed and filtered (even though it has already been processed prior to distribution through interstate pipelines) to ensure low moisture and particulate content. In addition, mercaptan - the source of the familiar rotten egg smell in natural gas - is added by the utility prior to distribution. This is added because natural gas is odorless and colorless, and the familiar odor of mercaptan makes the detection of leaks much easier.

Traditionally, rigid steel pipe was used to construct distribution networks. However, new technology is allowing the use of flexible plastic and corrugated stainless steel tubing in place of rigid steel pipe. These new types of tubing allow cost reduction, installation flexibility and easier repairs for both local distribution companies and natural gas consumers.

Another innovation in the distribution of natural gas is the use of electronic meter-reading systems. The natural gas that is consumed by any one customer is measured by on-site meters, which essentially keep track of the volume of natural gas consumed at that location. Traditionally, in order to bill customers correctly, meter-reading personnel had to be dispatched to record these volumes. However, new electronic meter-reading systems are capable of transmitting this information directly to the utility. This results in cost savings for the utility, which are in turn passed along to customers. The installation of natural gas distribution pipe requires the same process as for larger pipelines: the excavation of trenches into which the pipe is laid. However, new trenching techniques are allowing for the installation of distribution pipe with less impact on the above ground surroundings. Guided drilling systems are used to excavate an underground hole in which the pipe may be inserted, and can lead to significant excavation and restoration savings. This is particularly important in crowded urban settings and scenic rural environments, where the installation of natural gas distribution pipe can be a major inconvenience for residents and business owners.

Supervisory control and data acquisition (SCADA) systems, similar to those used by large pipeline companies, are also used by local distribution companies. These systems can integrate gas flow control and measurement with other accounting, billing, and contract systems to provide a comprehensive measurement and control system for the local gas utility. This allows accurate, timely information on the status of the distribution network to be used by the utility, to ensure efficient and effective service at all times.

Regulation of Natural Gas Distribution

Traditionally, local gas utilities have been awarded exclusive rights to distribute natural gas in a specified geographic area, as well as perform services like billing, safety inspection, and providing natural gas hookups for new customers. Like interstate pipelines, utilities have historically been viewed as natural monopolies. Because of the high cost of constructing the distribution infrastructure, it is uneconomic to lay multiple redundant distribution networks in any one area, resulting in only one utility offering distribution services. Because of their position as natural monopolies in a given geographic area, distribution companies have historically been regulated to ensure that monopoly power is not abused, and natural gas consumers do not fall victim to overly high distribution costs or inefficient delivery systems.

State public utility commissions are charged with the oversight and regulation of investor owned local natural gas utilities. Those utilities owned by local governments are typically governed by local government agencies to ensure that the needs and preferences of customers are met in a cost effective manner.

State regulation of local distribution companies has a variety of objectives, including ensuring adequate supply, dependable service, and reasonable prices for consumers, while also allowing for an adequate rate of return for investor owned Utilities. State regulators are also responsible for overseeing the construction of new distribution networks, including approving installation sites and proposed additions to the network. Regulatory orders and methods of oversight vary from state to state.

Historically, local distribution companies offered only "bundled" services; that is, they combined the cost of transportation, distribution, and the natural gas itself into one price for consumers. However, beginning in the 1990s, residential "customer choice" programs began to be offered as part of a movement toward the retail "unbundling" of natural gas sales. Many states now offer programs in which customers may choose a supplier from whom to purchase the natural gas commodity separately, and use the gas utility simply for service and delivery of that gas. Customer choice programs are in place in more than 20 states and the District of Columbia. To learn more about the status of state distribution customer choice programs, visit EIA.

Although the majority of residential and small commercial customers still tend to purchase 'bundled' natural gas from utilities, the increasingly important role of natural gas marketers, as well as the innovation fuelled by increasing competition in the marketplace, is leading to innovative ways of supplying natural gas to small volume users as well as new bundled service options, such as home security systems. Please visit our section to learn more about natural gas marketing in residential markets.

Distribution and Safety

Local distribution companies, like the larger interstate and intrastate pipelines, maintain the highest safety standards to ensure that preventable accidents are avoided, and problems with the distribution network are remedied in a timely fashion. Many of the safety programs maintained by utilities are quite similar to those of interstate pipeline companies. Safety measures at the local level include:

- Leak Detection Equipment – Utilities have in place sophisticated leak detection equipment, designed to pick up on leaks of natural gas from the distribution network. Utilities also add odorants to the natural gas to make it easier to detect a leak.
- Safety Education Programs - Utilities typically run natural gas safety seminars in schools, community centers, and through other organizations to ensure customers are well versed in

natural gas safety procedures and know what to do in the event of a leak or emergency.

- Technicians on Call - Utilities maintain fleets of technicians on call 24 hours a day, seven days a week to respond to customers' problems and concerns.
- Emergency Preparedness - Utilities participate in community and local emergency preparedness programs, educating and preparing for emergency events such as natural disasters.
- One Call Systems - Provides customers, contractors, and excavators with a single phone number to call before commencing excavation or construction, to ensure that the pipelines, and other buried facilities are not damaged. A national "call-before-you-dig" phone number of "811" was adopted in 2008 with the support of utilities, communities, emergency responders and government officials.

These are but a few of the safety measures maintained by local distribution companies. Especially important for the safe distribution of natural gas, particularly in densely populated areas, is the education of customers. By teaching natural gas users the safe use of natural gas, what to do in an emergency, and how to detect leaks, distribution companies ensure that the distribution of natural gas will remain one of the safest forms of energy transmission.

Marketing

Natural gas marketing is a relatively new addition to the natural gas industry, beginning in the mid-1980's. Prior to the deregulation of the natural gas commodity market and the introduction of open access for everyone to natural gas pipelines, there was no role for natural gas marketers.

Producers sold to pipelines, who sold to local distribution companies and other large volume natural gas users. Local distribution companies sold the natural gas purchased from the pipelines to retail end users, including commercial and residential customers. Price regulation at all levels of this supply chain left no place for others to buy and sell natural gas.

However, with the newly accessible competitive markets introduced gradually over the past fifteen years, natural gas marketing has become an integral component of the natural gas industry. In fact, the first marketers were a direct result of interstate pipelines attempting to recoup losses associated with long term contracts entered into as a result of the oversupply problems of the early 1980s. Natural gas marketing may be defined as the selling of natural gas. In even looser terms, marketing can be referred to as the process of coordinating, at various levels, the business of bringing natural gas from the wellhead to end-users.

The role of natural gas marketers is quite complex, and does not fit exactly into any one spot in the natural gas supply chain. Marketers may be affiliates of producers, pipelines, and local utilities, or may be separate business entities unaffiliated with any other players in the natural gas industry.

Marketers, in whatever form, find buyers for natural gas, ensure secure supplies of natural gas in the market, and provide a pathway for natural gas to reach the end-user. It is natural gas marketers that ensure a liquid, transparent market exists for natural gas. Marketing natural gas can include all of the intermediate steps that a particular purchase requires; including arranging transportation, storage, accounting, and basically any other step required to facilitate the sale of natural gas.

Essentially, marketers are primarily concerned with selling natural gas, either to resellers (other marketers and distribution companies), or end users.

On average, most natural gas can have three to four separate owners before it actually reaches the end-user. In addition to the buying and selling of natural gas, marketers use their expertise in financial instruments and markets to both reduce their exposure to risks inherent to commodities, and earn money through speculating as to future market movements.

In order to more fully understand the role and function of natural gas marketers, it is helpful to have an understanding of the basics of natural gas markets.

Natural Gas as a Commodity

Natural gas is sold as a commodity, much like pork bellies, corn, copper, and oil. The basic characteristic of a commodity is that it is essentially the same product no matter where it is located. Natural gas, after processing, fits this description.

Commodity markets are inherently volatile, meaning the price of commodities can change often, and at times drastically. Natural gas is no exception; in fact, it is one of the most volatile commodities currently on the market. The graph below shows the

The price of natural gas is set by market forces; the buying and selling of the commodity by market players, based on supply and demand, determines the average price of natural gas. There are two distinct markets for natural gas: the spot market, and the futures market. Essentially, the spot market is the daily market, where natural gas is bought and sold 'right now'. To get the price of natural gas on a specific day, it is the spot market price that is most informative. The futures market consists of buying and selling natural gas under contract at least one month, and up to 36 months, in advance.

For example, under a simplified futures contract, one could enter into an agreement today, for delivery of the physical gas in two months. Natural gas futures are traded on the New York Mercantile Exchange (NYMEX). Futures contracts are but one of an increasing number of derivatives contracts used in commodities markets, and can be quite complex and difficult to understand.

Natural gas is priced and traded at different locations throughout the country. These locations, referred to as 'market hubs', exist across the country and are located at the intersection of major pipeline systems. There are over 30 major market hubs in the U.S., the principle of which is known as the Henry Hub, located in Louisiana. The futures contracts that are traded on the NYMEX are Henry Hub contracts, meaning they reflect the price of natural gas for physical delivery at this hub.

The price at which natural gas trades differs across the major hubs, depending on the supply and demand for natural gas at that particular point. The difference between the Henry Hub price and another hub is called the location differential. In addition to market hubs, other major pricing locations include 'citygates'. Citygates are the locations at which distribution companies receive gas from a pipeline. Citygates at major metropolitan centers can offer another point at which natural gas is priced.

Physical and Financial Trading

There are two primary types of natural gas marketing and trading: physical trading and financial trading. Physical natural gas marketing is the more basic type, which involves buying and selling the physical commodity. Financial trading, on the other hand, involves derivatives and sophisticated financial instruments in which the buyer and seller never take physical delivery of the natural gas.

Like all commodity markets, the inherent volatility of the price of natural gas requires the use of financial derivatives to hedge against the risk of price movement. Buyers and sellers of natural gas hedge using derivatives to reduce price risk. Speculators, on the other hand, assume greater risk in order to profit off of changes in the price of natural gas.

Some marketers who actively buy and sell in either the physical or financial markets are referred to as natural gas 'traders'; trading natural gas on the spot market to earn as high a return as possible, and trading financial derivatives and other complex contracts to either hedge risk associated with this physical trading, or speculate about market movements. Most marketing companies have elaborate trading floors, including televisions and pricing boards providing the traders with as much market information as possible.

Physical Contracts

Physical trading contracts are negotiated between buyers and sellers. There exist numerous types of physical trading contracts, but most share some standard specifications including specifying the buyer and seller, the price, the amount of natural gas to be sold (usually expressed in a volume per day), the receipt and delivery point, the tenure of the contract (usually expressed in number of days, beginning on a specified day), and other terms and conditions. The special terms and conditions usually outline such things as the payment dates, quality specifications for the natural gas to be sold, and any other specifications agreed to by both parties.

Physical contracts are usually negotiated between buyers and sellers over the phone. However, electronic bulletin boards and e-commerce trading sites are allowing more physical transactions to take place over the internet.

There are three main types of physical trading contracts: swing contracts, baseload contracts, and firm contracts. Swing (or 'interruptible') contracts are usually short-term contracts, and can be as short as one day and are usually not longer than a month. Under this type of contract, both the buyer and seller agree that neither party is obligated to deliver or receive the exact volume specified. These contracts are the most flexible, and are usually put in place when either the supply of gas from the seller, or the demand for gas from the buyer, are unreliable.

Baseload contracts are similar to swing contracts. Neither the buyer nor seller is obligated to deliver or receive the exact volume specified. However, it is agreed that both parties will attempt to deliver or receive the specified volume, on a 'best-efforts' basis. In addition, both parties generally agree not to end the agreement due to market price movements. Both of these understandings are not legal obligations - there is no legal recourse for either party if they believe the other party did not make its best effort to fulfil the agreement - they rely instead on the relationship (both personal and professional) between the buyer and seller.

Firm contracts are different from swing and baseload contracts in that there is legal recourse available to either party, should the other party fail to meet its obligations under the agreement. This means that both parties are legally obligated to either receive or deliver the amount of gas specified in the contract. These contracts are used primarily when both the supply and demand for the specified amount of natural gas are unlikely to change or drop off.

The daily spot market for natural gas is active, and trading can occur 24 hours a day, seven days a week. However, in the natural gas market, the largest volume of trading occurs in the last week of every month. Known as 'bid week', this is when producers are trying to sell their core production and consumers are trying to buy for their core natural gas needs for the upcoming month.

The core natural gas supply or demand is not expected to change; producers know they will have that much natural gas over the next month, and consumers know that they will require that much natural gas over the next month. The average prices set during bid week are commonly the prices used in physical contracts.

The Financial Market

In addition to trading physical natural gas, there is a significant market for natural gas derivatives and financial instruments in the United States. In fact, it has been estimated that the value of trading that occurs on the financial market is 10 to 12 times greater than the value of physical natural gas trading.

Derivatives are financial instruments that 'derive' their value from an underlying fundamental; in this case the price of natural gas. Derivatives can range from being quite simple, to being exceedingly complex. Traditionally, most derivatives are traded on the over-the-counter (OTC) market, which is essentially a group of market players interested in exchanging certain derivatives among themselves, as opposed to through a market like the NYMEX. Basic types of derivatives include futures, options, and financial swaps.

There are two possible objectives to trading in financial natural gas markets: hedging and speculation. Trading in the physical market involves a certain degree of risk. Price volatility in the natural gas markets can result in financial exposure for marketers and other market players as the price changes over time. Trading financial derivatives can help to mitigate, or 'hedge' this risk. A hedging strategy is created to reduce the risk of losing money. Purchasing homeowner's insurance is a common hedging activity.

Similarly, a marketer who plans on selling natural gas in the spot market for the next month may be worried about falling prices, and can use a variety of financial instruments to hedge against the possibility of natural gas being worth less in the future. Countless strategies exist to hedge against price risk in the natural gas market, including natural gas futures, derivatives based on weather conditions to mitigate the risk of weather affecting the supply of natural gas (and thus its market price), etc. To learn more about the basics of hedging in the natural gas market visit the New York Mercantile Exchange here. Financial natural gas markets may also be used by market participants who wish to speculate about price movements or related events that may come about in the future. The main difference between speculation and hedging is that the objective of hedging is to reduce risk, whereas the objective of speculation is to take on risk in the hope of earning a financial return. Speculators hope to forecast future events or price movements correctly, and profit through these forecasts using financial derivatives.

Trading in the financial markets for speculative purpose is essentially making an investment in financial markets tied to natural gas, and financial speculators need not have any vested interest in the buying or selling of natural gas itself, only in the inherent underlying value that is represented in financial derivatives.

While great profits may be made if the expectations of a speculator prove correct, great losses may also be incurred if these expectations are wrong. While the instruments used for hedging and speculation are the same, the way in which they are used determines whether or not they in fact reduce, or increase, the risk of losing money.

Now that some of the basics of the natural gas market have been covered, we can examine the function of natural gas marketers.

Natural Gas Marketers

Any party who engages in the sale of natural gas can be termed a marketer, however they are usually specialized business entities dedicated solely to transacting in the physical and financial energy markets. It is commonplace for natural gas marketers to be active in a number of energy markets, taking advantage of their knowledge of these markets to diversify their business. Many natural gas marketers are also involved in the marketing of electricity, and in certain instances crude oil.

Marketers can be producers of natural gas, pipeline marketing affiliates, distribution utility marketing affiliates, independent marketers, and large volume users of natural gas. A recent study of the origins of natural gas marketers found that 27 percent of the top 30 natural gas marketers in 2000 were entities spun off from interstate pipeline companies. An equal percentage was made up of entities affiliated with local distribution companies.

About 30 percent of the top natural gas marketers were originally affiliated with producers, and entities formed from large volume natural gas consumers comprise 6 percent. Finally, independent, newly formed entities represent 10 percent of top natural gas marketers.

Marketing companies, whether affiliated with another member of the natural gas industry or not, can vary in size and the scope of their operations. Some marketing companies may offer a full range of services, marketing numerous forms of energy and financial products, while others may be more limited in their scope.

For instance, most marketing firms affiliated with producers do not sell natural gas from third parties; they are more concerned with selling their own production, and hedging to protect their profit margin from these sales.

There are basically five different classifications of marketing companies: major nationally integrated marketers, producer marketers, small geographically focused marketers, aggregators, and brokers.

The major nationally integrated marketers are the 'big players', offering a full range of services, and marketing numerous different products. They operate on a nationwide basis, and have large amounts of capital to support their trading and marketing operations. Producer marketers are those entities generally concerned with selling their own natural gas production, or the production of their affiliated natural gas production company.

Smaller marketers target particular geographic areas, and specific natural gas markets. Many marketing entities affiliated with LDCs are of this type, focusing on marketing gas for the geographic area in which their affiliated distributor operates. Aggregators generally gather small volumes from various sources, combine them, and sell the larger volumes for more favourable prices and terms than would be possible selling the smaller volumes separately. Brokers are a unique class of marketers in that they never actually take ownership of any natural gas themselves. They simply act as facilitators, bringing buyers and sellers of natural gas together.

All marketing companies must have, in addition to the core trading group, significant 'backroom' operations. These support staff are responsible for coordinating everything related to the sale and purchase of physical and financial natural gas; including arranging transportation and storage, posting completed transactions, billing, accounting, and any other activity that is required to complete the purchases and sales arranged by the traders. Since marketers generally work with very slim profit margins, the efficiency and effectiveness of these backroom operations can make a large impact on the profitability of the entire marketing operation.

In addition to the traders and backroom staff, marketing companies typically have extensive risk management operations. The risk management team is responsible for ensuring that the traders do not expose the marketing company to excessive risk. Top-level management is responsible for setting guidelines and risk limitations for the marketing operations, and it is up to the risk management team to ensure that traders comply with these directives. Risk management operations are quite complex, and rely on complex statistical, mathematical, and financial theory to ensure that risk exposure is kept under control. Most large losses associated with marketing operations occur when risk management policies are ignored or are not enforced within the company itself.

The marketing of natural gas is an integral part of the natural gas supply chain. Natural gas marketers ensure that a viable market for natural gas exists at all times. Efficient and effective physical and financial markets are the only way to ensure that a fair and equitable commodity price, reflective of the supply and demand for that commodity, is maintained.

Environment & Technology

The preservation of our environment is a very important and pressing topic, particularly when dealing with energy issues. The advancement of technology, particularly technologies that allow the cleaner use of fossil fuels, may provide many environmental benefits and allow us to use cleaner energy. This section focuses on the environmental issues related to the use of natural gas, as well as the advancement of new and exciting technologies within the industry.

- The Natural Gas and the Environment section describes the environmental effects of the use of natural gas, including comparing the emissions from natural gas to other fossil fuels, and describes efforts on the part of the natural gas industry to lessen the environmental impact of natural gas production.
- The Natural Gas and Technology section focuses on the advancement of technology in the natural gas industry, and includes resources for exploring new technologies in the industry.

Natural Gas and Technology

Over the past few decades the oil and natural gas industry has transformed into one of the most technologically advanced industries in the United States. New innovations have reshaped the industry into a technology leader. This section will discuss the role of technology in the evolution of the natural gas industry, focusing on technologies in the exploration and production sector, as well as a few select innovations that have had a profound effect on the potential for natural gas. Scroll down, or the links below to jump ahead:

- Advances in the Exploration and Production Sector
- Liquefied Natural Gas
- Natural Gas Fuel Cells
- Natural Gas Technology Resources.

Beginning in the 1990s, more power generators began to use natural gas to make electricity, causing demand for natural gas to grow substantially. The natural gas industry has been able to keep pace with growing demand and produce greater amounts of natural gas through technological innovation. These innovations have enabled the development of natural gas from shale and other formerly "unconventional" formations that are found in abundance across the United States, as well as development from traditional offshore and onshore formations. Below is a brief list of some of the major technological advancements that have been made recently:

Advances in the Exploration and Production Sector

Technological innovation in the exploration and production sector has equipped the industry with the equipment and practices necessary to continually increase the production of natural gas to meet rising demand. These technologies serve to make the exploration and production of natural gas more efficient, safe, and environmentally friendly. Even as natural gas deposits are increasingly produced from "unconventional" formations such as shale rock, the exploration and production industry has not only kept up its production pace, but in fact has improved the general nature of its operations, contributing to an unprecedented 39 percent increase in the size of U.S. resources since 2006.

According to a Department of Energy Report, "*Environmental Benefits of Advanced Oil and Gas Exploration and Production Technology,*" released in 1999 and still one of the most indepth analyses available as of 2010:

- 22,000 fewer wells are needed on an annual basis to develop the same amount of oil and gas reserves as were developed in 1985.
- Had technology remained constant since 1985, it would take two wells to produce the same amount of oil and natural gas as one 1985 well. However, advances in technology mean that one well today can produce two times as much as a single 1985 well.
- Drilling wastes have decreased by as much as 148 million barrels due to increased well productivity and fewer wells.
- The drilling footprint of well pads has decreased by as much as 70 percent due to advanced drilling technology, which is extremely useful for drilling in sensitive areas.
- By using modular drilling rigs and slimhole drilling, the size and weight of drilling rigs can be reduced by up to 75 percent over traditional drilling rigs, reducing their surface impact.
- Had technology, and thus drilling footprints, remained at 1985 levels, today's drilling footprints would take up an additional 17,000 acres of land.
- New exploration techniques and vibrational sources mean less reliance on explosives, reducing the impact of exploration on the environment.

New exploration techniques and vibrational sources mean less reliance on explosives, reducing the impact of exploration on the environment. New exploration techniques and vibrational sources mean less reliance on explosives, reducing the impact of exploration on the environment.

Some of the major recent technological innovations in the exploration and production sector include:

- 3-D and 4-D Seismic Imaging - The development of seismic imaging in three dimensions greatly changed the nature of natural gas exploration. This technology uses traditional seismic imaging techniques, combined with powerful computers and processors, to create a three-dimensional model of the subsurface layers. 4-D seismology expands on this, by adding time as a dimension, allowing exploration teams to observe how subsurface characteristics change over time. Exploration teams can now identify natural gas prospects more easily, place wells more effectively, reduce the number of dry holes drilled, reduce drilling costs, and cut exploration time. This leads to both economic and environmental benefits.
- CO2-Sand Fracturing - Fracturing techniques have been used since the 1970s to help increase the flow rate of natural gas and oil from underground formations. CO2-Sand fracturing involves using a mixture of sand proppants and liquid CO2 to fracture formations, creating and enlarging cracks through which oil and natural gas may flow more freely. The CO2 then vaporizes, leaving only sand in the formation, holding the newly enlarged cracks open. Because there are no other substances used in this type of fracturing, there are no 'leftovers' from the fracturing process that must be removed. This means that, while this type of fracturing effectively opens the formation and allows for increased recovery of oil and natural gas, it does not damage the deposit, generates no below ground wastes, and protects groundwater resources.
- Coiled Tubing - Coiled tubing technologies replace the traditional rigid, jointed drill pipe with a long, flexible coiled pipe string. This greatly reduces the cost of drilling, as well as providing a smaller drilling footprint, requiring less drilling mud, faster rig set up, and reducing the time normally needed to make drill pipe connections. Coiled tubing can also be used in combination with slimhole drilling to provide very economic drilling conditions, and less impact on the environment.
- Measurement While Drilling - Measurement-While-Drilling (MWD) systems allow for the collection of data from the bottom of a well as it is being drilled. This allows engineers and drilling teams access to up-to-the-second information on the exact nature of the rock formations being encountered by the drill

bit. This improves drilling efficiency and accuracy in the drilling process, allows better formation evaluation as the drill bit encounters the underground formation, and reduces the chance of formation damage and blowouts.

- Slimhole Drilling - Slimhole drilling is exactly as it sounds; drilling a slimmer hole in the ground to get to natural gas and oil deposits. In order to be considered slimhole drilling, at least 90 percent of a well must be drilled with a drill bit less than six inches in diameter (whereas conventional wells typically use drill bits as large as 12.25 inches in diameter). Slimhole drilling can significantly improve the efficiency of drilling operations, as well as decrease its environmental impact. In fact, shorter drilling times and smaller drilling crews can translate into a 50 percent reduction in drilling costs, while reducing the drilling footprint by as much as 75 percent. Because of its low cost profile and reduced environmental impact, slimhole drilling provides a method of economically drilling exploratory wells in new areas, drilling deeper wells in existing fields, and providing an efficient means for extracting more natural gas and oil from un-depleted fields.
- Offshore Drilling Technology - The offshore oil and natural gas production sector is sometimes compared to the aeronautics field and NASA due to achievements in deepwater drilling that have been facilitated by state of the art technology. New technology, including improved offshore drilling rigs, dynamic positioning devices and sophisticated navigation systems are allowing safe, efficient offshore drilling in waters more than 10,000 feet deep. Visit the offshore drilling section to learn more.
- Hydraulic Fracturing also called "Fracking," or "Frac'ing"- Used to free natural gas that is trapped in shale rock formations. A liquid mix that is 99 percent water and sand is injected into the rock at very high pressure, creating fractures within the rock that provide the natural gas a path to flow to the wellhead. The fracking fluid mix also helps to keep the formation more porous. Hydraulic fracturing is now widely used, with more than 90 percent of the natural gas wells in the United States having used it to boost production at some time.

The above technological advancements provide only a snapshot of the increasingly sophisticated technology being developed and put into practice in the exploration and production of natural gas and oil. New technologies and applications are being developed constantly, and serve to improve the economics of producing natural gas, allow for the production of deposits formerly considered too unconventional or uneconomic to develop, and ensure that the supply of natural gas keeps up with steadily increasing demand. Sufficient domestic natural gas resources exist to help fuel the U.S. for a significant period of time, and technology is playing a huge role in providing low-cost, environmentally sound methods of extracting these resources.

Liquefied Natural Gas

Cooling natural gas to about -260°F at normal pressure results in the condensation of the gas into liquid form, known as Liquefied Natural Gas (LNG). LNG can be very useful, particularly for the transportation of natural gas, since LNG takes up about one six hundredth the volume of gaseous natural gas. Advances in technology are reducing the costs associated with the liquification and regasification of LNG. Because it is easy to transport, LNG can serve to make economical stranded natural gas deposits from around the globe for which the construction of pipelines is uneconomical.

LNG, when vaporized to gaseous form, will only burn in concentrations of between 5 and 15 percent mixed with air. In addition, LNG, or any vapour associated with LNG, will not explode in an unconfined environment. Thus, in the unlikely event of an LNG spill, the natural gas has little chance of igniting an explosion. Liquification removes oxygen, carbon dioxide, sulfur, and water from the natural gas, resulting in LNG that is almost pure methane.

LNG is typically transported by specialized tanker with insulated walls, and is kept in liquid form by autorefrigeration, a process in which the LNG is kept at its boiling point, so that any heat additions are countered by the energy lost from LNG vapour that is vented out of storage and used to power the vessel.

The increased use of LNG is allowing for the production and marketing of natural gas deposits that were previously economically unrecoverable. Although it currently accounts for only about 1 percent of natural gas used in the United States, it is expected that LNG imports will provide a steady, dependable source of natural gas for U.S. consumption.

Natural Gas Fuel Cells

Fuel cells powered by natural gas are an extremely exciting and promising new technology for the clean and efficient generation of electricity. Fuel cells have the ability to generate electricity using electrochemical reactions as opposed to combustion of fossil fuels to generate electricity. Essentially, a fuel cell works by passing streams of fuel (usually hydrogen) and oxidants over electrodes that are separated by an electrolyte. This produces a chemical reaction that generates electricity without requiring the combustion of fuel, or the addition of heat as is common in the traditional generation of electricity. When pure hydrogen is used as fuel, and pure oxygen is used as the oxidant, the reaction that takes place within a fuel cell produces only water, heat, and electricity. In practice, fuel cells result in very low emission of harmful pollutants, and the generation of high-quality, reliable electricity. The use of natural gas-powered fuel cells has a number of benefits, including:

- *Clean Electricity* - Fuel cells provide the cleanest method of producing electricity from fossil fuels. While a pure hydrogen, pure oxygen fuel cell produces only water, electricity, and heat, fuel cells in practice emit trace amounts of sulfur compounds and very low levels of carbon dioxide. However, the carbon dioxide produced by fuel cell use is concentrated and can be readily recaptured, as opposed to being emitted into the atmosphere.
- *Distributed Generation* - Fuel cells can come in extremely compact sizes, allowing for their placement wherever electricity is needed. This includes residential, commercial, industrial, and even transportation settings.
- *Dependability* - Fuel cells are completely enclosed units, with no moving parts or complicated machinery. This translates into a dependable source of electricity, capable of operating for thousands of hours. In addition, they are very quiet and safe sources of electricity. Fuel cells also do not have electricity surges, meaning they can be used where a constant, dependable source of electricity is needed.
- *Efficiency* - Fuel cells convert the energy stored within fossil fuels into electricity much more efficiently than traditional generation of electricity using combustion. This means that less fuel is required to produce the same amount of electricity. The National Energy Technology Laboratory estimates that, used in combination with natural gas turbines, fuel cell generation facilities can be produced that will operate in the 1 to 20 Megawatt range at 70 percent efficiency, which is much higher than the efficiencies that can be reached by traditional generation methods within that output range.

The generation of electricity has traditionally been a very polluting, inefficient process. However, with new fuel cell technology, the future of electricity generation is expected to change dramatically in the next ten to twenty years. Research and development into fuel cell technology is ongoing, to ensure that the technology is refined to a level where it is cost-effective for all varieties of electric generation requirements.

Natural Gas Technology Resources

The natural gas industry is joined by government agencies and laboratories, private research and development firms, and environmental technology groups in coming up with new technologies that may improve the efficiency, cost-effectiveness, and environmental soundness of the natural gas industry. New technology and methods emerge frequently in the natural gas industry. Below are links to a number of resources that provide information on new technological developments in the oil and natural gas industry.

Chapter 5

Environmental Impact of the Oil Shale Industry

Environmental impact of the oil shale industry includes the consideration of issues such as land use, waste management, and water and air pollution caused by the extraction and processing of oil shale. Surface mining of oil shale deposits causes the usual environmental impacts of open-pit mining.

In addition, the combustion and thermal processing generate waste material, which must be disposed of, and harmful atmospheric emissions, including carbon dioxide, a major greenhouse gas. Experimental in-situ conversion processes and carbon capture and storage technologies may reduce some of these concerns in future, but may raise others, such as the pollution of groundwater.

Surface Mining and Retorting

Land use and Waste Management

Surface mining and *in-situ* processing requires extensive land use. Mining, processing and waste disposal require land to be withdrawn from traditional uses, and therefore should avoid high density population areas. Oil shale mining reduces the original ecosystem diversity with habitats supporting a variety of plants and animals.

After mining the land has to be reclaimed. However, this process takes time and cannot necessarily re-establish the original biodiversity. The impact of sub-surface mining on the surroundings will be less than for open pit mines. However, sub-surface mining may also cause subsidence of the surface due to the collapse of mined-out area and abandoned stone drifts.

Disposal of mining wastes, spent oil shale (including semi-coke) and combustion ashes needs additional land use. According to the study of the European Academies Science Advisory Council, after processing, the waste material occupies a greater volume than the material extracted, and therefore cannot be wholly disposed underground. According to this, production of a barrel of shale oil can generate up to 1.5 tons of semi-coke, which may occupy up to 25% greater volume than the original shale. This is not confirmed by the results of the Estonia's oil shale industry. The mining and processing of 1000 million tons of oil shale in Estonia has created about 360-370 million tons of solid waste, of which 90 million tons is a mining waste, 70–80 million tons is a semi-coke, and 200 million tons are combustion ashes.

The waste material may consist of several pollutants including sulfates, heavy metals, and polycylic aromatic hydrocarbons (PAHs), some of which are toxic and carcinogenic. To avoid contamination of the groundwater, the solid waste from the thermal treatment process is disposed in an open dump (landfill or "heaps"), not underground. As semi-coke consists of, in addition to minerals, up to 10 % organics that may pose hazard to the environment owing to leaching of toxic compounds as well as to the possibility of self-ignition.

Water Management

Mining influences the water runoff pattern of the area affected. In some cases it requires the lowering of groundwater levels below the level of the oil shale strata, which may have harmful effects on the surrounding arable land and forest. In Estonia, for each cubic meter of oil shale mined, 25 cubic meters of water must be pumped from the mine area. At the same time, the thermal processing of oil shale needs water for quenching hot products and the control of dust. Water concerns are particularly sensitive issue in arid regions, such as the western part of the United States and Israel's Negev Desert, where there are plans to expand the oil shale industry. Depending on technology, above-ground retorting uses between one and five barrels of water per barrel of produced shale oil. *In situ* processing, according to one estimate, uses about one-tenth as much water.

Water represents the major vector of transfer of oil shale industry pollutants. One environmental issue is to prevent noxious materials leaching from spent shale into the water supply. The oil shale processing is accompanied by the formation of process waters and waste waters containing phenols, tar and several other products, heavily separable and toxic to the environment. A 2008 programmatic environmental impact statement issued by the United States Bureau of Land Management stated that surface mining and retort operations produce 2 to 10 U.S. gallons (7.6 to 38 l; 1.7 to 8.3 imp gal) of waste water per 1 short ton (0.91 t) of processed oil shale.

Air Pollution Management

Main air pollution is caused by the oil shale-fired power plants, which provide the atmospheric emissions of gaseous products like nitrogen oxides, sulfur dioxide and hydrogen chloride, and the airborne particulate matter (fly ash). It includes particles of different types (carbonaceous, inorganic ones) and different sizes. The concentration of air pollutants in flue gas depends primarily on the combustion technology and burning regime, while the emissions of solid particles are determined by the efficiency of fly ash-capturing devices. Open deposition of semi-coke causes distribution of pollutants in addition to aqueous vectors also via air (dust).

Greenhouse Gas Emissions

Carbon dioxide emissions from the production of shale oil and shale gas are higher than conventional oil production and a report for the European Union warns that increasing public concern about the adverse consequences of global warming may lead to opposition to oil shale development.

Emissions arise from several sources. These include CO_2 released by the decomposition of the kerogen and carbonate minerals in the extraction process, the generation of the energy needed to heat the shale and in the other oil and gas processing operations, and fuel used in the mining of the rock and the disposal of waste. As the varying mineral composition and calorific value of oil shale deposits varies widely, the actual values vary considerably. At best, the direct combustion of oil shales produces carbon emissions similar to those from the lowest form of coal, lignite, at 2.15 moles CO_2/MJ, an energy source which is also politically contentious due to its high emission levels. For both power generation and oil extraction, the CO_2 emissions can be reduced by better utilization of waste heat from the product streams.

In-situ Processing

Currently, the *in-situ* process is the most attractive proposition due to the reduction in standard surface environmental problems. However, *in-situ* processes do involve possible significant environmental costs to aquifers, especially since *in-situ* methods may require ice-capping or some other form of barrier to restrict the flow of the newly gained oil into the groundwater aquifers. However, after the removal of the freeze wall these methods can still cause groundwater contamination as the hydraulic conductivity of the remaining shale increases allowing groundwater to flow through and leach salts from the newly toxic aquifer.

Oil Shale Geology

Oil shale geology is a branch of geologic sciences which studies the formation and composition of oil shales–fine-grained sedimentary rocks containing significant amounts of kerogen, and belonging to the group of sapropel fuels. Oil shale formation takes place in a number of depositional settings and has considerable compositional variation. Oil shales can be classified by their composition (carbonate minerals such as calcite or detrital minerals such as quartz and clays) or by their depositional environment (large lakes, shallow marine, and lagoon/ small lake settings). Much of the organic matter in oil shale is of algal origin, but may also include remains of vascular land plants. Three major type of organic matter (macerals) in oil shale are telalginite, lamalginite, and bituminite. Some oil-shale deposits also contain metals which include vanadium, zinc, copper, uranium.

Most oil shale deposits were formed during Middle Cambrian, Early and Middle Ordovician, Late Devonian, Late Jurassic, and Paleogene times through burial by sedimentary loading on top of the algal swamp deposits, resulting in conversion of the organic matter to kerogen by diagenetic processes. The largest deposits are found in the remains of large lakes such as the deposits of the Green River Formation of Wyoming and Utah, USA. Oil-shale deposits formed in the shallow seas of continental shelves generally are much thinner than large lake basin deposits.

Classification and Varieties

There are varying classifications of oil shales depending on their mineral content, type of kerogen, age, depositional history, and organisms from which they are derived. The age of the oil shale deposits ranges from Cambrian to Tertiary age. Lithologies range from shales to marl and carbonate rocks, all of which form a mixture of tightly bound organic and inorganic materials.

Oil shales have been divided into three categories based on mineral composition – carbonate-rich shale, siliceous shale and cannel shale. Carbonate-rich shales derive their name from the large amount of carbonate minerals such as calcite and dolomite. As many as twenty carbonate minerals have been found in oil shale, the majority of which are considered authigenic or diagentic. Carbonate-rich oil shales, particularly that of lacustrine-sourced deposits, have usually the organic-rich layers sandwiched between carbonate-rich layers.

These deposits are hard formations that are resistant to weathering and they are difficult to process using *ex-situ* methods. Siliceous oil shales are usually dark brown or black shales. They are not rich in carbonates but rather in siliceous minerals such as quartz, feldspar, clay, chert and opal. Siliceous shales are not as hard and weather-resistant as carbonate-rich shales, and may be better suited for extraction via *ex-situ* methods. Cannel shales are usually dark brown or black shales, which consist of organic matter that completely encloses other mineral grains. They are suitable for extraction via *ex-situ* methods.

Another classification according to the type of kerogen, is based on the hydrogen, carbon, and oxygen content of oil shales' original organic matter. This classification is known as the "van Krevelen diagram". The most used classification of oil shales was developed between 1987 and 1991 by Adrian C. Hutton of the University of Wollongong, adapting petrographic terms from coal terminology. According to this classification, oil shales are designated as terrestrial, lacustrine (lake-bottom-deposited), or marine (ocean bottom-deposited), based on the environment where the initial biomass was deposited. Hutton's classification scheme has proven useful in estimating the yield and composition of the extracted oil.

Table: *Classification of oil shales by environment of deposition*

Terrestrial	***Lacustrine***	***Marine***
cannel coal	lamosite;torbanite	kukersite; tasmanite; marinite

Cannel coal (also called *candle coal*) is a type of terrestrial shale, which is hydrogen-rich brown to black coal, sometimes with shaly texture, composed of resins, spores, waxes, cutinaceous and corky materials derived from terrestrial vascular plants as well as varied amounts of vitrinite and inertinite. Lacustrine shales consist of Lamosite and Torbanite.

Lamosite is a pale-brown and grayish-brown to dark-gray to black oil shale whose chief organic constituent is lamalginite derived from lacustrine planktonic algae. Torbanite, named after Torbane Hill in

Scotland, is a black oil shale whose organic matter is telalginite derived from lipid-rich *Botryococcus* and related algal forms. Marine shales consist of three varieties, namely Kukersite, Tasmanite, and Marinite. Kukersite, named after Kukruse in Estonia, is a light-brown marine oil shale whose principal organic component is telalginite derived from the green alga, *Gloeocapsomorpha prisca*.

Tasmanite, named after Tasmania, is a brown to black oil shale whose organic matter consists of telalginite derived chiefly from unicellular tasmanitid algae of marine origin. Marinite is a gray to dark-gray to black oil shale of marine origin in which the chief organic components are lamalginite and bituminite derived from marine phytoplankton with varied admixtures of bitumen, telalginite, and vitrinite.

Composition

Photomicrograph showing detail of the varves in a rich Colorado oil shale specimen. The organic laminae are themselves finely laminated. The mineral laminae contain considerable organic matter, but they are readily distinguished by their coarser grain and greater thickness. Note sand grains (white). Enlarged 320 diameters.

As a sapropel fuel, oil shale differs from humus fuels in its lower content of organic matter. The organic matter has an atomic ratio of hydrogen to carbon of about 1.5 – approximately the same as that of crude oil and four to five times higher than coals. The organic matter in oil shales forms a complex macromolecular structure which is insoluble in common organic solvents. It is mixed with varied amounts of mineral matter. For commercial grades of oil shale, the ratio of organic matter to mineral matter is about 0.75:5 to 1.5:5.

The organic portion of oil shale consists largely of prebitumen bituminous groundmass, such as remains of algae, spores, pollen, plant cuticles and corky fragments of herbaceous and woody plants, and cellular debris from other lacustrine, marine, and land plants. While terrestrial oil shales contain resins, spores, waxy cuticles, and corky tissues of roots and stems of vascular terrestrial plants, lacustrine oil shales include lipid-rich organic matter derived from algae. Marine oil shales are composed of marine algae, acritarchs, and marine dinoflagellates. Organic matter in oil shale also contains organic sulfur (about 1.8 % on average) and a lower proportion of nitrogen.

Three major types of organic matter (macerals) in oil shale are telalginite, lamalginite, and bituminite. Telalginite is defined as structured organic matter composed of large colonial or thick-walled unicellular algae such as Botryococcus and *Tasmanites*. Lamalginite includes thin-walled colonial or unicellular algae that occur as distinct laminae, but display few or no recognizable biologic structures.

Under the microscope, telalginite and lamalginite are easily recognized by their bright shades of yellow under ultraviolet/blue fluorescent light. Bituminite is largely amorphous, lacks recognizable biologic structures, and displays relatively low fluorescence under the microscope. Other organic constituents include vitrinite and inertinite, which are macerals derived from the humic matter of land plants. These macerals are usually found in relatively small amounts in most oil shales.

Formation

Most oil shale formations took place during mid-Cambrian, early and middle Ordovician, late Devonian, late Jurassic and Paleogene periods. These were formed by the deposition of organic matter in a variety of depositional environments including freshwater to highly saline lakes, epicontinental marine basins and subtidal shelves and were restricted to estuarine areas such as oxbow lakes, peat bogs, limnic and coastal swamps, and muskegs. When plants die in such an anaerobic aquatic environment, low oxygen levels prevent their complete bacterial decay.

For undecayed organic matter to be preserved and to form oil shale, the environment must remain uniform for prolonged periods of time in order to build up sufficiently thick sequences of algal matter. Eventually, the algal swamp or other restricted environment is disrupted and oil shale accumulation ceases. Burial by sedimentary loading on top of the algal swamp deposits converts the organic matter to kerogen by the following normal diagenetic processes:

- Compaction due to sediment loading on the coal, leading to compression of the organic matter.
- With ongoing heat and compaction, removal of moisture in the peat and from the intracellular structure of fossilized plants, and removal of molecular water.
- Methanogenesis—similar to treating wood in a pressure cooker— results in methane being produced, removing hydrogen, some carbon, and some further oxygen.
- Dehydration, which removes hydroxyl groups from the cellulose and other plant molecules, resulting in the production of hydrogen-reduced coals or oil shales.

Though similar in their formation process, oil shales differ from coals in several distinct ways. The precursors of the organic matter in oil shale and coal differ in a sense that oil shale is of algal origin, but may also include remains of vascular land plants that more commonly compose much of the organic matter in coal.

The origin of some of the organic matter in oil shale is obscure because of the lack of recognizable biologic structures that would help identify the precursor organisms. Such materials may be of bacterial origin or the product of bacterial degradation of algae or other organic matter. Lower temperature and pressure during the diagenesis process compared to other modes of hydrocarbon generation result in a lower maturation level of oil shale. Continuous burial and further heating and pressure could result in the production of oil and gas from the oil shale source rock.

The largest deposits are found in the remains of large lakes such as the deposits of the Green River Formation of Wyoming and Utah, USA. Large lake oil shale basins are typically found in areas of block faulting or crustal warping due to mountain building. Deposits such as the Green River may be as much as 2,000 feet (610 m) and yield up to 40 gallons of oil for each ton (166 l/t) of shale.

Oil-shale deposits formed in the shallow seas of continental shelves generally are much thinner than large lake basin deposits. These are typically a few meters thick and are spread over very large areas, extending up to thousands of square kilometres. Of the three lithologic types of oil shales, siliceous oil shales are most commonly found in such environment. These oil shales are not as organically rich as lake-deposited oil shales, and generally do not contain more than 30 gallons per ton of oil shale. Oil shales deposited in lagoonal or small lake environments are rarely extensive and are often associated with coal-bearing rocks. These oil shales can have high yields– as much as 40 gallons per ton (166 l/t) of oil shale. However, due to their small areal extent, they are considered unlikely candidates for commercial exploitation.

Formations in the United States

The United States has two significant oil-shale deposits which are suited for commercial development due to their size, grade and location. The Eocene Green River Formation covers parts of Colorado, Wyoming and Utah; the second significant deposit is Devonian oil shales in the eastern United States. In both places, there are sub-basins varying in volume and quality of the reserves. Oil shale in the Green River

Formation is found in five sedimentary basins namely, Green River, Uinta, Piceance Creek, Sand Wash and Washakie. The first three have undergone some significant exploration and attempts to commercialize the oil shale reserves since the 1960s. The Green River Formation includes deposits from two large lakes which covered an area of over 65,000 square kilometres (25,100 sq mi) during early to middle Eocene period. These lakes were separated by the Uinta uplift and the Axial Basin anticline.

For significant periods during their 10 Ma life, the lakes became closed systems allowing many changes in size, salinity and sediment deposition. Oil shale is a result of abundant blue-green algae that thrived in the lakes.

The oil shale that underlies almost 750,000 square kilometres (289,580 sq mi) in the eastern United States was formed in a marine depositional environment very different from the Green River Basins. These deposits have also undergone commercialization attempts; they are also resources for natural gas and have been mined for low-grade oil shale. These oil shales were formed during the Late Devonian and Early Mississippian periods. During this time, much of the eastern United States was covered by a large shallow sea. The oil shale is thought to have been the result of slow deposition of planktonic algae. under anoxic conditions. In parts of the basin close to the shoreline, the organic mixture that helped form the oil shale contains organic-rich sediment from the rising Appalachian mountains.

Formations in Brazil

Brazil has nine significant locations of oil shale deposits. The size, location and quality of oil shale deposits in the Paraiba Valley and the Irati Formation have attracted the most attention. These two contain an estimated 1.4 billion barrels of *in-situ* shale oil with total resources as much as more than three billion barrels. While the "Irati formation" deposit is the smaller of the two, containing an estimated 600 million barrels *in-situ* compared to 840 million in the Paraiba valley formation, the former is more economically viable.

The "Irati formation" consists of two oil shale beds separated by 12 metres (40 ft) of limestone and shale. The upper layer is thicker (9 metres (30 ft)) but the thinner lower bed (4 metres (10 ft)) is of greater value; the weight percent of shale oil yield is around 12 % for the lower layer as compared to 7 % for the upper one. The oil shale yield varies laterally, and may be as little as 7 % for the lower layer and 4 % for the upper layer. The formation is a very fine grained and laminated deposit ranging in colour from dark gray to brown to black. While 60-70 % of the shale consists of clay minerals, the balance is made up of organic matter.

No consensus has been reached on the exact depositional nature of the Irati oil shale. One theory suggests that the organic material in the Irati oil shale originated from algae deposited in a lacustrine environment with salinity varying from that of freshwater to brackish water. Other theory suggests that the organic sediment may have been deposited in a shallow, partially restricted marine environment. Hutton's classification describes it as a marine source oil shale.

Formation in Estonia

The kukersite oil shale of Ordovician age in Estonia is part of the Baltic oil shale basin and was deposited in shallow marine basins. The deposit is one of the world's highest-grade deposits with more than 40 % organic content and 66 % conversion ratio into shale oil and gas. The oil shale is located in a single calcareous layer 2.5-3 meters in thickness and is buried at depths from 7 to 100 m. The total area of the basin is about 3,000 km^2. Oil yield from Kukersite is 30 to 47 %. Most of the organic matter is derived from the fossil green alga, *Gloeocapsomorpha prisca*, which has affinities to the modern cyanobacterium, *Entophysalis major*, an extant species that forms algal mats in inter-tidal to very shallow subtidal waters. Matrix minerals include low-magnesium calcite, dolomite, and siliciclastic minerals. It is not enriched in heavy metals.

Oil Shale Reserves

Oil shale reserves refers to oil shale resources that are recoverable under given economic restraints and technological abilities. Oil shale deposits range from small presently non-economic occurrences to large presently commercially exploitable reserves. Defining oil shale reserves is difficult, as the chemical composition of different oil shales, as well as their kerogen content and extraction technologies, vary significantly. The economic feasibility of shale oil extraction is highly dependent on the price of conventional oil; if the price of crude oil per barrel is less than the production price per barrel of shale oil, it is uneconomic.

There are around 600 known oil shale deposits. Many deposits need more exploration to determine their potential as reserves. However, worldwide technically recoverable reserves have recently been estimated at about 2.8–3.3 trillion barrels (450×10^9–520×10^9 m^3)

of shale oil, with the largest reserves in the United States, which is thought to have 1.5–2.6 trillion barrels (240×10^9–410×10^9 m^3). Well-explored deposits, which could be classified as reserves, include the Green River deposits in the western United States, the Tertiary deposits in Queensland, Australia, deposits in Sweden and Estonia, the El-Lajjun deposit in Jordan, and deposits in France, Germany, Brazil, China, and Russia. It is expected that these deposits would yield at least 40 liters (0.25 bbl) of shale oil per metric ton of shale, using the Fischer Assay.

Definition of Reserves

Estimating shale oil reserves is complicated by several factors. Firstly, the amount of kerogen contained in oil shale deposits varies considerably. Secondly, some nations report as reserves the total amount of kerogen in place, including all kerogen regardless of technical or economic constraints; these estimates do not consider the amount of kerogen that may be extracted from identified and assayed oil shale rock using available technology and under given economic conditions. By most definitions, "reserves" refers only to the amount of resource which is technically exploitable and economically feasible under current economic conditions.

The term "resources", on the other hand, may refer to all deposits containing kerogen. Thirdly, shale oil extraction technologies are still developing, so the amount of recoverable kerogen can only be estimated.

There are a wide variety of extraction methods, which yield significantly different quantities of useful oil. As a result, the estimated amounts of resources and reserves display wide variance. The kerogen content of oil shale formations differs widely, and the economic feasibility of its extraction is highly dependent on international and local costs of oil. Several methods are used to determine the quantity and quality of the products extracted from shale oil. At their best, these methods give an approximate value to its energy potential. One standard method is the Fischer Assay, which yields a heating value, that is, a measure of caloric output. This is generally considered a good overall measure of usefulness. The Fischer Assay has been modified, standardized, and adapted by the American Petroleum Institute. It does not, however, indicate how much oil could be extracted from the sample. Some processing methods yield considerably more useful product than the Fischer Assay would indicate. The Tosco II method yields over 100% more oil, and the Hytort process yields between 300% to 400% more oil.

Table: *Largest oil shale deposits (over 1 billion metric tons)*

Deposit	***Country***	***Period***	***In-place shale oil resources (million barrels)***	***In-place oil shale resources (million metric tons)***
Green River Formation	USA	Tertiary	1,466,000	213,000
Phosphoria Formation	USA	Permian	250,000	35,775
Eastern Devonian	USA	Devonian	189,000	27,000
Heath Formation	USA	Early Carboniferous	180,000	25,578
Olenyok Basin	Russia	Cambrian	167,715	24,000
Congo	Democratic Republic of Congo	?	100,000	14,310
Irati Formation	Brazil	Permian	80,000	11,448
Sicily	Italy	?	63,000	9,015
Tarfaya	Morocco	Cretaceous	42,145	6,448
Volga Basin	Russia	?	31,447	4,500
St. Petersburg, Baltic Oil Shale Basin	Russia	Ordovician	25,157	3,600
Vychegodsk Basin	Russia	Jurassic	19,580	2,800
Wadi Maghar	Jordan	Cretaceous	14,009	2,149
Dictyonema shale	Estonia	Ordovician	12,386	1,900
Timahdit	Morocco	Cretaceous	11,236	1,719
Collingwood Shale	Canada	Ordovician	12,300	1,717
Italy	Italy	Triassic	10,000	1,431

Geographical Allocation

There is no comprehensive overview of oil shales geographical allocation around the world. Around 600 known oil shale deposits are diversely spread throughout the earth, and are found on every continent with the possible exception of Antarctica, which has not yet been explored for oil shale.

Oil shale resources can be concentrated in a large confined deposit such as the Green River formations, which were formed by a large inland lake. These can be many meters thick but limited by the size of the original lake. They may also resemble the deposits found along the eastern American seaboard, which were the product of a shallow sea, in that they may be quite thin but laterally expansive, covering thousands of square kilometres.

The table below reports reserves by estimated amount of shale oil. Shale oil refers to synthetic oil obtained by heating organic material (kerogen) contained in oil shale to a temperature which will separate it into oil, combustible gas, and the residual carbon that remains in the spent shale. All figures are presented in barrels and metric tons.

Shale oil: resources and production at end-2005 by regions and countries with resources over 10 billion barrels (1.6 km^3) of in-place shale oil

Region	*In-place shale oil resources (million barrels)*	*In-place oil shale resources (million metric tons)*	*Production in 2002 (thousand metric tons (oil))*
Africa	159,243	23,317	-
Democratic Republic of the Congo	100,000	14,310	-
Morocco	53,381	8,187	-
Asia	45,894	6,562	180
China	16,000	2,290	180
Europe	368,156	52,845	345
Russia	247,883	35,470	-
Italy	73,000	10,446	-
Estonia	16,286	2,494	345
Middle East	38,172	5,792	-
Jordan	34,172	5,242	-
North America	2,100,469	303,758	-
United States	2,085,228	301,566	-
Canada	15,241	2,192	-
Oceania	31,748	4,534	-
Australia	31,729	4,531	-
South America	82,421	11,794	157
Brazil	82,000	11,734	159
World total	***2,826,103***	***408,602***	***684***

Africa

Major oil shale deposits are located in the Democratic Republic of Congo (equal to 14.31 billion metric tons of shale oil) and Morocco (12.3 billion metric tons or 8.16 billion metric tons of shale oil). Deposits in Congo are not properly explored yet. In Morocco, oil shale deposits have been identified at ten localities with the largest deposits in Tarfaya and Timahdite. Although reserves in Tarfaya and Timahdit are well explored, the commercial exploitation has not started yet and only a limited program of laboratory and pilot-plant research has been undertaken. There are also oil shale reserves in Egypt, South Africa, Madagascar, and Nigeria. The main deposits of Egypt are located in Safaga-Al-Qusair and Abu Tartour areas.

Asia

Major oil shale deposits are located in China, which has an estimated total of 32 billion metric tons, of which 4.4 billion metric tons are technically exploitable and economically feasible; Thailand (18.7 billion metric tons), Kazakhstan (several deposits; major deposit at Kenderlyk Field with 4 billion metric tons), and Turkey (2.2 billion metric tons). Smaller oil shale reserves have also been found in Assam (India), Pakistan, Uzbekistan, Turkmenistan, Myanmar, Armenia, and Mongolia.

The principal Chinese oil shale deposits and production lie in Fushun and Liaoning; others are located in Maoming in Guangdong, Huadian in Jilin, Heilongjiang, and Shandong. In 2002, China produced more than 90,000 metric tons of shale oil. Thailand's oil shale deposits are near Mae Sot, Tak Province, and at Li, Lamphun Province. Deposits in Turkey are found mainly in middle and western Anatolia.

Professor Alan R. Carroll of University of Wisconsin–Madison estimates that Upper Permian lacustrine oil shale deposits of northwest China, absent from previous global oil shale assessments, are comparable to the Green River Formation.

Europe

The biggest oil shale reserves in Europe are located in Russia (equal to 35.47 billion metric tons of shale oil). Major deposits are located in the Volga-Petchyorsk province and in the Baltic Oil Shale Basin. Other major oil shale deposits in Europe are located in Italy (10.45 billion metric tons of shale oil), Estonia (2.49 billion metric tons of shale oil), France (1 billion metric tons of shale oil), Belarus (1 billion metric tons of shale oil), Sweden (875 million metric tons of shale oil), Ukraine (600 million metric tons of shale oil) and the United Kingdom

(500 million metric tons of shale oil). There are oil shale reserves also in Germany, Luxembourg, Spain, Bulgaria, Hungary, Poland, Serbia, Austria, Albania, and Romania.

Middle East

Significant oil shale deposits are located in Jordan (5.242 billion metric tons of shale oil or 65 billion metric tons of oil shale) and Israel (550 million metric tons of shale oil or 6.5 billion metric tons of oil shale). Jordanian oil shales are high quality, comparable to western US oil shale, although their sulfur content is high. The best-explored deposits are El Lajjun, Sultani, and the Juref ed Darawishare located in west-central Jordan, while the Yarmouk deposit, close to its northern border, extends into Syria. Most of Israel's deposits are located in the Rotem Basin region of the northern Negev desert near the Dead Sea. Israeli oil shale is relatively low in heating value and oil yield.

North America

At 301 billion metric tons, the oil shale deposits in the United States are easily the largest in the world. There are two major deposits: the eastern US deposits, in Devonian-Mississippian shales, cover 250,000 square miles (650,000 km^2); the western US deposits of the Green River Formation in Colorado, Wyoming, and Utah, are among the richest oil shale deposits in the world. In Canada 19 deposits have been identified. The best-examined deposits are in Nova Scotia and New Brunswick.

Oceania

Australia's oil shale resource is estimated at about 58 billion metric tons or 4.531 billion metric tons of shale oil, of which about 24 billion barrels (3.8 billion cubic metres) is recoverable. The deposits are located in the eastern and southern states with the biggest potential in the eastern Queensland deposits. Oil shale has also been found in New Zealand.

South America

Brazil has the world's second-largest known oil shale resources (the Irati shale and lacustrine deposits) and is currently the world's second largest shale oil producer, after Estonia. Oil shale resources occur in São Mateus do Sul, Paraná, and in Vale do Paraíba. Brazil has developed the world's largest surface oil shale pyrolysis retort at Petrosix, with a 11-meter (36 ft)-diameter vertical shaft. Brazilian production in 1999 was about 200,000 metric tons. Small resources are also found in Argentina, Chile, Paraguay, Peru, Uruguay, and Venezuela.

Alberta Taciuk Process

The Alberta Taciuk process (ATP; known also as the AOSTRA Taciuk process) is an above-ground dry thermal retorting technology for extracting oil from oil sands, oil shale and other organics-bearing materials, including oil contaminated soils, sludges and wastes. The technology is named after its inventor William Taciuk and the Alberta Oil Sands Technology and Research Authority.

History

The research and development of the ATP technology started in 1970. In 1975, its inventor, William Taciuk, formed the UMATAC Industrial Processes (now part of Polysius) to further its development. The first ATP pilot plant was constructed in 1977.

The ATP was originally developed for pyrolysis of oil sand. However, its first commercial application in 1989 was dedicated to the environmental remediation of contaminated soils. From 1999 to 2004, ATP technology was used for shale oil extraction at the Stuart Oil Shale Plant in Australia. During that time, 1.5 million barrels (238.48094×10^3 m^3) of shale oil was extracted before the owner, Southern Pacific Petroleum Pty Ltd went into receivership. The subsequent owner, Queensland Energy Resources closed and dismantled the plant.

In 2002, Estonian company Viru Keemia Grupp tested this technology; however, it was not taken into use.

Technology

The ATP is an above-ground oil-shale retorting technology classified as a hot recycled solids technology. The distinguishing feature of the ATP is that the drying and pyrolysis of the oil shale or other feed, as well as the combustion, recycling, and cooling of spent materials and residues, all occur within a single rotating multi-chamber horizontal retort. Its feed consists of fine particles.

In its shale-oil applications, fine particles (less than 25 millimetres (1.0 in) in diameter) are fed into the preheat tubes of the retort, where they are dried and preheated to 250 °C (480 °F) indirectly by hot shale ash and hot flue gas. In the pyrolysis zone, oil shale particles are mixed with hot shale ash and the pyrolysis is performed at temperatures between 500 °C (930 °F) and 550 °C (1,020 °F). The resulting shale oil vapour is withdrawn from the retort through a vapour tube and recovered by condensation in other equipment. The char residues, mixed with ash, are moved to the combustion zone, and burnt at about 800 °C (1,470 °F) to form shale ash. Part of the ash is delivered to the pyrolysis zone, where its heat is recycled as a hot solid carrier; the other part is removed and cooled in the cooling zone with the combustion gases by heat transfer to the feed oil shale.

The advantages of the ATP technology for shale oil extraction lie in its simple and robust design, energy self-sufficiency, minimal process water requirements, ability to handle fine particles, and high oil yields. It is particularly suited for processing materials with otherwise low oil yield. The mechanical transfer of solids through the machine does not involve moving parts and it achieves improved process efficiencies through solid-to-solid heat transfer. Most of the process energy (over 80%) is produced by combustion of char and produced oil shale gas; external energy inputs are minimal. The oil yields are about 85–90% of Fischer Assay. The organic carbon content of the process residue (spent shale) is less than 3%. The process produces only small amounts of contaminated water with low concentrations of phenols. These advantages also apply to its oil sands applications, including increased oil yield, a simplified process flow, reduction of bitumen losses to tailings, elimination of the need for tailing ponds, improvement in energy efficiency compared with the hot water extraction process, and elimination of requirements for chemical and other additives.

A complication of the ATP is that retorting operations can reach temperatures at which carbonate minerals within the shale decompose, increasing greenhouse gas emissions.

Operations

As of 2008, ATP was used by the United States Environmental Protection Agency at a PCB-contaminated site near Buffalo, New York, and at the Waukegan Harbor, Illinois.

UMATAC Industrial Processes runs a 5 tons of oil shale per hour pilot processor in Calgary, Alberta for large scale tests of different oil shales. The Fushun Mining Group of China has built a 250 tonnes per hour ATP plant that began commissioning in 2010. Jordan Energy and Mining Ltd plans to use the ATP technology for extracting oil from Al Lajjun and Attarat oil shale deposits in Jordan.

Fischer Assay

The Fischer assay is a standardized laboratory test for determining the oil yield from oil shale to be expected from a conventional shale oil extraction. A 100 gram oil shale sample crushed to <2.38 mm is heated in a small aluminum retort to 500 °C (930 °F) at a rate of 12°C/min (22°F/min), and held at that temperature for 40 minutes. The distilled vapours of oil, gas, and water are passed through a condenser and cooled with ice water into a graduated centrifuge tube. The oil yields achieved by other technologies are often reported as a percentage of the Fischer Assay oil yield.

The original Fischer Assay test was developed in the early low temperature coal retorting research by Franz Joseph Emil Fischer and Hans Schrader. It was adapted for evaluating oil shale yields in 1949 by K. E. Stanfield and I. C. Frost.

Fushun Process

The Fushun process is an above-ground retorting technology for shale oil extraction. It is named after the main production site of Fushun, Liaoning province in northeastern China.

History

The Fushun process was developed and utilized for the extraction of shale oil in China during the mid-1920s. The commercial-scale utilization of the process began in 1930 with the construction of "Refinery No. 1". After World War II, the shale oil production was ceased, but 100 Fushun-type oil shale retorts were restored in 1949. In 1950, total 266 retorts were in operation, each with the capacity of 100–200 tons of shale oil per day.

With the discovery of Daqing oil field in 1960s, the shale oil production declined and Sinopec, an operator of shale oil production these times, shut down its oil shale operations in the beginning of 1990s. At the same time, the Fushun Oil Shale Retorting Plant, using Fushun process technology, was established as a part of the Fushun Mining Group. It started production in 1992. In 2005, China became the largest shale oil producer in the world. In 1985-1986, Sinopec used the Fushun process for a test processing of Jordan's oil shale from the El Lajjun deposit. Although the process was technically viable, the cooperation was halted due to high operation costs.

Technology

The Fushun process is classified as an internal combustion technology but also includes external gas heating. It uses a vertical cylindrical type shaft retort, with outside steel plate lined with inner fire bricks. The retort has height over 10 metres (33 ft) and its inner diameter is about 3 metres (9.8 ft). Raw oil shale particles with the size of 10 to 75 millimetres (0.4 to 3.0 in) are fed from the top of the retort. At the upper section of the retort oil shale is dried and heated by the ascending hot gases, which pass upward through the descending oil shale causing decomposition of the rock. The pyrolysis takes place at about 500 °C (930 °F). The produced oil vapour and gases exit from the top of the retort; hot gases and oil vapours move from the bottom to the top directly, and not diagonally like in Kiviter process. During the pyrolysis process, oil shale is decomposed to shale coke (char), which together with the ascending air-steam is burnt in the lower part of the retort to heat gases necessary for pyrolysis. These gases are recirculated; after leaving retort, they are cooled in a condensation system, where shale oil is condensed, and re-heated in a heating furnace about 500 °C (930 °F) to 700 °C (1,290 °F) before reinserting into the retort. The shale ash exits from a rotating water dish that acts as a seal and cooler at the bottom of the retort.

Retorts are operated in sets and have a heat carrier preparation unit and rotating water seals designed for the whole set instead of a single retort as in case of the Kiviter retort. Regenerative furnaces are located next to the retorts and they are operating in two cycles – the combustion cycle and the gas heating cycle. During the combustion cycle, a furnace is heated up to 1,000 °C (1,830 °F) by combustion gases. After the combustion cycle, retort gases from the condensation system is inserted into a furnace for their heating. By alternating furnaces, one furnace is always available for heating retort gas. Twenty retorts typically share one condensation system and a set of heating furnances.

Advantages of the Fushun process include small investment and stable operation. The process is characterized by the high thermal efficiency, but due to the addition of air into the retort, the nitrogen dilutes the pyrolysis gas. In addition, the excess oxygen in retort burns out a part of produced shale oil, which reduce the shale oil yield. The oil yield of the Fushun retort accounts for about 65% of Fischer Assay. Disadvantage of this process is a high water consumption amounting to 6–7 barrels of water per barrel of produced shale-oil, and great quantities of waste shale. It is not suitable of ores with small size and oil content lower than 5%.

As the capacity of single retort is limited, Fushun process is suitable for small-scale retorting plants, and for processing lean oil shale with low gas yield.

Commercial Use

Fushun process is used only in China. The Fushun Mining Group operates the largest by capacity shale oil plant in the world (Fushun Shale Oil Plant) consisting of 180 Fushun retorts. Each retort processes about 4 tonnes of oil shale per hour.

Galoter Process

The Galoter process (also known as TSK, UTT, or SHC; its newest modifications called Enefit and Petroter) is a shale oil extraction technology for a production of shale oil, a type of synthetic crude oil. In this process, the oil shale is decomposed into shale oil, oil shale gas, and spent residue. A decomposition is caused by mixing raw oil shale with a hot oil shale ash, generated by combustion of carbonaceous residue (semi-coke) in the spent residue. The process was developed in 1950s and it is used commercially for the shale oil production in Estonia. There are projects for further development of this technology and for expansion of its usage, e.g. in Jordan and USA.

History

Research of the solid heat carrier process for pyrolysis of lignite, peat, and oil shale started in 1944 at the G. M. Krzhizhanovsky Power Engineering Institute of Academy of Sciences of the USSR. At the laboratory scale, the Galoter process was invented and developed in 1945–1946. The process was named Galoter after the research team leader I. Galinker who's name was combined with the word "thermal".

The further research continued in Estonia. A pilot unit with capacity of 2.5 tonnes of oil shale per day was built in Tallinn in 1947. The first Galoter-type commercial scale pilot retorts were built at Kiviõli, Estonia, in 1953 and 1963 (closed in 1963 and 1981 respectively) with capacities of 200 and 500 tonnes of oil shale per day respectively. The Narva Oil Plant, annexed to the Eesti Power Plant and operating two Galoter-type 3000 tonnes per day retorts, was commissioned in Estonia in 1980. These retorts were designed by AtomEnergoProject and developed in cooperation with the Krzhizhanovsky Institute. Started as a pilot plant, the process of converting it to commercial scale plant took about 20 years. During this period, the company has modernized more than 70% of the equipment compared to the initial design.

In 1978, a 12.5-tonnes pilot plant was built in Verkhne-Sinevidnoy, Ukraine. It was used for testing of Lviv–Volinsk lignite, and Carpathian, Kashpir (Russia) and Rotem (Israel) oil shales. In 1996–1997, a test unit was assembled in Tver.

In 2008, Estonian energy company Eesti Energia, an operator of Galoter retorts at the Narva Oil Plant, established a joint venture with the Finnish technology company Outotec called Enefit Outotec Technology to develop and commercialize a modified Galoter process–the Enefit process–which combines the current process with circulating fluidized bed technologies.

Technology

Galoter Retort

The Galoter process is an above-ground oil-shale retorting technology classified as a hot recycled solids technology. The process uses a horizontal cylindrical rotating kiln-type retort, which is slightly declined. It has similarities with the TOSCO II process.

Before retorting, the oil shale is crushed into fine particles with a size of less than 25 millimetres (1.0 in) in diameter. The crushed oil shale is dried in the fluidized bed drier (aerofountain drier) by contact with hot gases. After drying and pre-heating to 135 °C (275 °F), oil shale particles are separated from gases by cyclonic separation. Oil shale is transported to the mixer chamber, where it is mixed with hot ash of 800 °C (1,470 °F), produced by combustion of spent oil shale in a separate furnace. The ratio of oil shale ash to raw oil shale is 2.8–3:1. The mixture is moved then to the hermetic rotating kiln. When the heat transfers from the hot ash to raw oil shale particles, the pyrolysis (chemical decomposition) begins in oxygen deficit conditions. The temperature of pyrolysis is kept at 520 °C (970 °F). Produced oil vapours and gases are cleaned of solids by cyclones and moved to condensation system (rectification column) where shale oil condense and oil shale gas is separated in gaseous form. Spent shale (semi-coke) is transported then to the separate furnace for combustion to produce hot ash. A portion of the hot ash is separated from the furnace gas by cyclones and recycled to the rotary kiln for pyrolysis. The remaining ash is removed from the combustion gas by more cyclones and cooled and removed for disposal by using water. The cleaned hot gas returns to the oil shale dryer.

The Galoter process has high thermal and technological efficiency, and high oil recovery ratio. Oil yield reaches 85–90% of Fischer Assay and retort gas yield accounts for 48 cubic meters per tonne. Oil quality is considered good, but the equipment is sophisticated and capacity is relatively low. This process creates less pollution than internal combustion technologies, as it uses less water, but it still generates carbon dioxide as also carbon disulfide and calcium sulfide.

Enefit Process

Enefit process is a modification of the Galoter process being developed by Enefit Outotec Technology. In this process, the Galoter technology is combined with proven circulating fluidized bed (CFB) combustion technology used in coal-fired power plants and mineral processing. Oil shale particles and hot oil shale ash are mixed in a rotary drum as in the classical Galoter process. The primary modification is the replacing of the Galoter semi-coke furnace with a CFB furnace. The Enefit process also incorporates fluid bed ash cooler and waste heat boiler commonly used in coal-fired boilers to convert waste heat to steam for power generation. Compared to the traditional Galoter, the Enefit process allows complete combustion of carbonaceous residue, improved energy efficiency by maximum utilization of waste heat, and less water use for quenching. According to promoters, the Enefit process has a lower retorting time compare to the classical Galoter process and therefore its has a greater throughput. Avoidance of moving parts in the retorting zones increases their durability.

Commercial Use

Two Galoter retorts built in 1980 are used for oil production by the Narva Oil Plant, a subsidiary of the Estonian energy company Eesti Energia. Both retorts process 125 tonnes per hour of oil shale. The annual shale oil production is 135,000 tonnes and oil shale gas production is 40 million cubic metres per annum (1.4×10^9 cu ft/a). The company is building a new plant employing its newer Enefit process with a processing capacity of 2.26 million tonnes of oil shale per year and producing 290,000 tonnes of shale oil and 75 million cubic metres (2.6×10^9 cu ft) of oil shale gas. It is planned to become operational by 2012. In addition, Eesti Energia plans to begin construction of a similar Enefit plants in Jordan by 2015 and in USA by 2017.

VKG Oil, a subsidiary of Viru Keemia Grupp, constructed a new Galoter retort (Petroter) at Kohtla-Järve that is similar to the two retorts operating at Narva. The basic engineering of the retort was done by Atomenergoproject of Saint Petersburg. The basic engineering of the condensation and distillation plant was done by Rintekno of Finland. The plant has a processing capacity of 1.1 million tonnes of oil shale per year and it produces 100,000 tonnes of shale oil, 30 million cubic metres (1.1×10^9 cu ft) of oil shale gas, and 150 GWh of steam per year. Operation started in December 2009. The company plans to start construction of the second Petroter plant in 2012.

Ex Situ Technologies

Internal Combustion

Internal combustion technologies burn materials (typically char and oil shale gas) within a vertical shaft retort to supply heat for pyrolysis. Typically raw oil shale particles between 12 millimetres (0.5 in) and 75 millimetres (3.0 in) in size are fed into the top of the retort and are heated by the rising hot gases, which pass through the descending oil shale, thereby causing decomposition of the kerogen at about 500 °C (932 °F). Shale oil mist, evolved gases and cooled combustion gases are removed from the top of the retort then moved to separation equipment. Condensed shale oil is collected, while non-condensable gas is recycled and used to carry heat up the retort. In the lower part of the retort, air is injected for the combustion which heats the spent oil shale and gases to between 700 °C (1,292 °F) and 900 °C (1,650 °F). Cold recycled gas may enter the bottom of the retort to cool the shale ash. The Union A and Superior Direct processes depart from this pattern. In the Union A process, oil shale is fed through the bottom of the retort and a pump moves it upward. In the Superior Direct process, oil shale is processed in a horizontal, segmented, doughnut-shaped travelling-grate retort.

Internal combustion technologies such as the Paraho Direct are thermally efficient, since combustion of char on the spent shale and heat recovered from the shale ash and evolved gases can provide all the heat requirements of the retort. These technologies can achieve 80-90% of Fischer assay yield. Two well-established shale oil industries use internal combustion technologies: Kiviter process facilities have been operated continuously in Estonia since the 1920s, and a number of Chinese companies operate Fushun process facilities.

Common drawbacks of internal combustion technologies are that the combustible oil shale gas is diluted by combustion gases and particles smaller than 10 millimetres (0.4 in) can not be processed. Uneven distribution of gas across the retort can result in blockages when hot spots cause particles to fuse or disintegrate.

Hot Recycled Solids

Hot recycled solids technologies deliver heat to the oil shale by recycling hot solid particles—typically oil shale ash. These technologies usually employ rotating kiln or fluidized bed retorts, fed by fine oil shale particles generally having a diameter of less than 10 millimetres (0.4 in); some technologies use particles even smaller than 2.5 millimetres (0.10 in). The recycled particles are heated in a separate chamber or vessel to about 800 °C (1,470 °F) and then mixed with the raw oil shale to cause the shale to decompose at about 500 °C (932 °F). Oil vapour and shale oil gas are separated from the solids and cooled to condense and collect the oil. Heat recovered from the combustion gases and shale ash may be used to dry and preheat the raw oil shale before it is mixed with the hot recycle solids.

In the Galoter and Enefit processes, the spent oil shale is burnt in a separate furnace and the resulting hot ash is separated from the combustion gas and mixed with oil shale particles in a rotating kiln. Combustion gases from the furnace are used to dry the oil shale in a dryer before mixing with hot ash. The TOSCO II process uses ceramic balls instead of shale ash as the hot recycled solids. The distinguishing feature of the Alberta Taciuk Process (ATP) is that the entire process occurs in a single rotating multi–chamber horizontal vessel.

Because the hot recycle solids are heated in a separate furnace, the oil shale gas from these technologies is not diluted with combustion exhaust gas. Another advantage is that there is no limit on the smallest particles that the retort can process, thus allowing all the crushed feed to be used. One disadvantge is that more water is used to handle the resulting finer shale ash.

Conduction Through a Wall

These technologies transfer heat to the oil shale by conducting it through the retort wall. The shale feed usually consists of fine particles. Their advantage lies in the fact that retort vapours are not combined with combustion exhaust. The Combustion Resources process uses a hydrogen–fired rotating kiln, where hot gas is circulated through an outer annulus. The Oil-Tech staged electrically heated retort consists of individual inter-connected heating chambers, stacked atop each other. Its principal advantage lies in its modular design, which enhances its portability and adaptability. The Red Leaf Resources EcoShale In-Capsule Process combines surface mining with a lower-temperature heating method similar to *in situ* processes by operating within the confines of an earthen structure. A hot gas circulated through parallel pipes heats the oil shale rubble. An installation within the empty space created by mining would permit rapid reclamation of the topography. A general drawback of conduction through a wall technologies is that the retorts are more costly when scaled-up due to the resulting large amount of heat conducting walls made of high-temperature alloys.

Externally Generated Hot Gas

In general, externally generated hot gas technologies are similar to internal combustion technologies in that they also process oil shale lumps in vertical shaft kilns. Significantly, though, the heat in these technologies is delivered by gases heated outside the retort vessel, and therefore the retort vapours are not diluted with combustion exhaust. The Petrosix and Paraho Indirect employ this technology. In addition to not accepting fine particles as feed, these technologies do not utilize the potential heat of combusting the char on the spent shale and thus must burn more valuable fuels. However, due to the lack of combustion of the spent shale, the oil shale does not exceed 500 °C (932 °F) and significant carbonate mineral decomposition and subsequent CO_2 generation can be avoided for some oil shales. Also, these technologies tend to be the more stable and easier to control than internal combustion or hot solid recyle technologies.

Reactive Fluids

Kerogen is tightly bound to the shale and resists dissolution by most solvents. Despite this constraint, extraction using especially reactive fluids has been tested, including those in a supercritical state. Reactive fluid technologies are suitable for processing oil shales with a low hydrogen content. In these technologies, hydrogen gas (H_2) or hydrogen donors (chemicals that donate hydrogen during chemical reactions) react with coke precursors (chemical structures in the oil shale that are prone to form char during retorting but have not yet done so). Reactive fluid technologies include the IGT Hytort (high-pressure H_2) process, donor solvent processes, and the Chattanooga fluidized bed reactor. In the IGT Hytort oil shale is processed in a high-pressure hydrogen environment. The Chattanooga process uses a fluidized bed reactor and an associated hydrogen-fired heater for oil shale thermal cracking and hydrogenation. Laboratory results indicate that these technologies can often obtain significantly higher oil yields than pyrolysis processes. Drawbacks are the additional cost and complexity of hydrogen production and high-pressure retort vessels.

In Situ Technologies

In situ technologies heat oil shale underground by injecting hot fluids into the rock formation, or by using linear or planar heating sources followed by thermal conduction and convection to distribute heat through the target area. Shale oil is then recovered through vertical wells drilled into the formation. These technologies are potentially able to extract more shale oil from a given area of land than conventional *ex situ* processing technologies, as the wells can reach greater depths than surface mines. They present an opportunity to recover shale oil from low-grade deposits that traditional mining techniques could not extract.

During World War II a modified *in situ* extraction process was implemented without significant success in Germany. One of the earliest successful *in situ* processes was underground gasification by electrical energy (Ljungström method)—a process exploited between

1940 and 1966 for shale oil extraction at Kvarntorp in Sweden. Prior to the 1980s, many variations of the *in situ* process were explored in the United States. The first modified *in situ* oil shale experiment in the United States was conducted by Occidental Petroleum in 1972 at Logan Wash, Colorado. The newest technologies explore a variety of heat sources and heat delivery systems.

Wall Conduction

Wall conduction *in situ* technologies use heating elements or heating pipes placed within the oil shale formation. The Shell in situ conversion process (Shell ICP) uses electrical heating elements for heating the oil shale layer to between 650 and 700 °F (340 and 370 °C) over a period of approximately four years. The processing area is isolated from surrounding groundwater by a freeze wall consisting of wells filled with a circulating super-chilled fluid. Disadvantages of this process are large electrical power consumption, extensive water use, and the risk of groundwater pollution. The process, under development since the early 1980s, is tested at the Mahogany test site in the Piceance Basin. 1,700 barrels (270 m^2) of oil were extracted in 2004 at a 30-by-40-foot (9.1 by 12 m) testing area.

In the American Shale Oil CCR Process, superheated steam or another heat transfer medium is circulated through a series of pipes placed below the oil shale layer to be extracted. The system combines horizontal wells, through which steam is passed, and vertical wells, which provide both vertical heat transfer through refluxing of converted shale oil and a means to collect the produced hydrocarbons. Heat is supplied by combustion of natural gas or propane in the initial phase and by oil shale gas at a later stage.

The Independent Energy Partners' Geothermic Fuels Cells Process (IEP GFC) extracts shale oil by exploiting a high-temperature stack of fuel cells. The cells, placed in the oil shale formation, are fuelled by natural gas during a warm-up period and afterward by oil shale gas generated by its own waste heat.

Externally Generated Hot Gas

Externally generated hot gas *in situ* technologies use hot gases that are heated above-ground and then injected into the oil shale formation. The Chevron CRUSH process, developed by Chevron Corporation in partnership with Los Alamos National Laboratory, injects heated carbon dioxide into the formation via drilled wells and heats the formation through a series of horizontal fractures in which the gas circulates. General Synfuels International has proposed the

Omnishale process which involves injecting super-heated air into the oil shale formation. Mountain West Energy's In Situ Vapour Extraction process uses similar principles of injection of high-temperature gas.

ExxonMobil Electrofrac

ExxonMobil's *in situ* technology (ExxonMobil Electrofrac) uses electrical heating with elements of both wall conduction and volumetric heating methods. It injects an electrically conductive material such as calcined petroleum coke into the hydraulic fractures created in the oil shale formation which then forms a heating element. Heating wells are placed in a parallel row with a second horizontal well intersecting them at their toe. This allows opposing electrical charges to be applied at either end.

Volumetric Heating

The Illinois Institute of Technology developed the concept of oil shale volumetric heating using radio waves (radio frequency processing) during the late 1970s. This technology was further developed by Lawrence Livermore National Laboratory. The oil shale would be heated by vertical electrode arrays. Deeper volumes could be processed at slower heating rates by installations spaced at tens of meters. The concept presumes a radio frequency at which the skin depth is many tens of meters, thereby overcoming the thermal diffusion times needed for conductive heating. Its drawbacks include intensive electrical demand and the possibility that groundwater or char would absorb undue amounts of the energy. Radio frequency processing in conjunction with critical fluids is being developed by Raytheon together with CF Technologies and tested by Schlumberger.

Microwave heating technologies are based on the same principles as radio wave heating, although it is believed that radio wave heating is an improvement over microwave heating because its energy can penetrate farther into the oil shale formation. The microwave heating process is being tested by Global Resource Corporation. Electro-Petroleum proposes electrically enhanced oil recovery by the passage of direct current between cathodes in producing wells and anodes located either at the surface or at depth in other wells. The passage of the current through the oil shale formation results in resistive Joule heating.

Oil Shale Economics

Oil shale economics deals with the economic feasibility of oil shale extraction and processing. The economic feasibility of oil shale is highly dependent on the price of conventional oil, and the assumption that the price will remain at a certain level for some time to come. As a developing fuel source the production and processing costs for oil shale are high due to the small nature of the projects and the specialist technology involved. A full-scale project to develop oil shale would require heavy investment and could potentially leave businesses vulnerable should the oil price drop, as the cost of producing the oil would exceed the price they could obtain for the oil.

Oil shale deposits in the USA, Estonia, China, and Brazil have been important over the past hundred years. At the current conditions few deposits can be exploited economically without subsidies. However, some countries, such as Estonia, Brazil, and China, operate oil-shale industries, while others, including Australia, USA, Canada, Jordan, Israel, and Egypt, are contemplating establishing or re-establishing this industry.

The production cost of a barrel of shale oil ranges from as high as US$95 per barrel to as low US$12 per barrel. The industry is proceeding cautiously, due to the losses incurred during the last major investment into oil shale in the early 1980s, when a subsequent collapse in the oil price left the projects uneconomical.

Competitive Level of Oil Price

The various attempts to develop oil shale deposits have succeeded only when the cost of shale-oil production in a given region comes in below the price of crude oil or its other substitutes. According to a survey conducted by the RAND Corporation, the cost of producing a barrel of oil at a surface retorting complex in the United States (comprising a mine, retorting plant, upgrading plant, supporting utilities, and spent shale reclamation), would range between US$70–95 ($440–600/m^3, adjusted to 2005 values). This estimate considers varying levels of kerogen quality and extraction efficiency. In order for the operation to be profitable, the price of crude oil would need to remain above these levels.

The analysis also discusses the expectation that processing costs would drop after the complex was established. The hypothetical unit would see a cost reduction of 35–70% after its first 500 million barrels (79×10^6 m^3) were produced. Assuming an increase in output of 25 thousand barrels per day (4.0×10^3 m^3/d) during each year after the start of commercial production, the costs would then be expected to decline to $35–48 per barrel ($220–300/m^3) within 12 years. After achieving the milestone of 1 billion barrels (160×10^6 m^3), its costs would decline further to \$30–40 per barrel (\$190–250/m^3). A comparison of the proposed American oil shale industry to the Alberta oil-sands industry has been drawn (the latter enterprise generated over 1 million barrels per day (160×10^3 m^3/d) of oil in late 2007), stating that "the first-generation facility is the hardest, both technically and economically".

In 2005, Royal Dutch Shell announced that its *in situ* extraction technology in Colorado could become competitive at prices over \$30 per barrel (\$190/m^3). However, Shell reported in 2007 that the cost of creating an underground freeze wall to contain groundwater contamination had significantly escalated.

At full-scale production, the production costs for one barrel of light crude oil of the Australia's Stuart plant were projected to be in the range of US\$11.3 to \$12.4 per barrel, including capital costs and operation costs over a projected 30-year lifetime. However, the project has been suspended due to environmental concerns.

The project of a new Alberta Taciuk Processor, planned by VKG Oil, is estimated to achieve break-even financial feasibility operating at 30% capacity, assuming a crude oil price of US\$21 per barrel or higher. At 50% utilization, the project is economic at a price of US\$18 per barrel, while at full capacity, it could be economic at a price of US\$13 per barrel.

Previous Investment

In the second half of the 20th century, oil shale production ceased in Canada, Scotland, Sweden, France, Australia, Romania, and South Africa due to the low price of oil and other competitive fuels. In the United States, during the 1973 oil crisis businesses expected oil prices to stay as high as US\$70 a barrel, and invested considerable sums in the oil shale industry. World production of oil shale reached a peak of 46 million tonnes in 1980. Due to competition from cheap conventional petroleum in the 1980s, several investments became economically unfeasible. On 2 May 1982, known as "*Black Sunday*", Exxon canceled its US\$5 billion Colony Shale Oil Project near Parachute, Colorado because of low oil-prices and increased expenses. Because of the losses in 1980s, companies were reluctant to make new invests in shale oil production. However, in the early 21st century, USA, Canada and Jordan were planning or had started shale oil production test projects, and Australia was considering restarting oil shale production.

In a 1972 publication by the journal *Pétrole Informations*, shale oil production was unfavorably compared to the liquefaction of coal.

The article stated that coal liquefaction was less expensive, generated more oil, and created fewer environmental impacts than oil shale extraction. It cited a conversion ration of 650 litres (170 U.S. gal; 140 imp gal) of oil per one ton of coal, as against 150 litres (40 U.S. gal) per one ton of shale oil.

Energy Usage

A measure of the viability of oil shale as a fuel source is the ratio of energy in the produced oil or electricity to the energy used in its conversion (Energy Returned on Energy Invested - EROEI). The value of the EROEI for oil shale is difficult to calculate for a number of reasons. Lack of reliable studies of modern oil shale processes, poor or undocumented methodology, and a limited number of operational facilities are the main reasons. Due to technically more complex process, the EROEI for oil shale is below the EROEI of about 20:1 for conventional oil extraction at the wellhead. Royal Dutch Shell has reported an EROEI about 3–4:1 on its in-situ development, Mahogany Research Project, which uses electric heating of the shale up to 500 °F (260 °C). A 1984 study estimated the EROEI of the different oil shale deposits to vary between 0.7–13.3:1. More recent studies estimates the EROEI of oil shales to be 1–2:1 or 2–16:1 – depending when self-energy is counted as a cost or internal energy is exclude and only purchased energy is counted as input. Self-energy is energy released by the oil shale conversion process that is used to power that operation (e.g. obtained by combustion of conversion by-products such as oil shale gas), and therefore reducing the use of other fuels (external energy).

There are different views if the internal energy should be added to the calculation as cost or not. One opinion is that internal energy should not be counted as an energy costs because is does not have an opportunity cost—unlike external energy used in the process. Other opinion is that internal energy is used for performing useful work and therefore should be added to the calculation. One argument is that this internal energy should be included as energy invested because it contributes to CO_2 emissions. However, EROEI then becomes a measure of environmental acceptability rather than economic viability.

Water Usage

Development of oil shale resources will require significant quantities of water for mine and plant operations, reclamation, supporting infrastructure, and associated economic growth. Above-ground retorting typically consumes between one and five barrels of water per barrel of produced shale oil, depending on technology. For an oil shale industry producing 2.5 million barrels per day (400×10^3 m^3/d), this equates to 105000000–315000000 US gallons per day (400,000–1,190,000 m^3/d) of water. These numbers include water requirements for power generation for in-situ heating processes, retorting, refining, reclamation, dust control and on-site worker demands. Municipal and other water requirements related to population growth associated with industry development will require an additional 58 million US gallons (220,000 m^3) per day. Hence, a 2.5 million barrels per day (400×10^3 m^3/d) oil shale industry would require 180,000 to 420,000 acre feet (220,000,000 to 520,000,000 m^3) of water per year, depending on location and processes used.

The largest deposit of oil shale in the United States is in the Green River basin. Though scarce, water in the western United States is treated as a commodity which can be bought and sold in a competitive market. Royal Dutch Shell has been reported to be buying groundwater rights in Colorado as it prepares to drill for oil in the shale deposits there. In the Colorado Big-Thompson project, average prices per share (0.7 acre feet (860 m^3)/share) increased from some $2,000 in 1990 to more than $12,000 in mid-2003 (constant 2001 dollars). CBT Prices from 2001 to 2006 has had a range of $10,000 to $14,000 per share, or $14,000 to $20,000 per acre foot. At $10,000 per acre foot, capital costs for water rights to produce 2.5 million barrels per day (400×10^3 m^3/d) would range between $1.8-4.2 billion.

Co-pyrolysis

Several co-pyrolysis processes to increase efficiency of oil shale retorting have been proposed or tested. In Estonia, the co-pyrolysis of kukersite with renewable fuel (wood waste), as well as with plastic and rubber wastes (tyres), has been tested. Co-pyrolysis of oil shale with high-density polyethylene (HDPE) has been tested also in Morocco and Turkey. Israel's AFSK Hom Tov co-pyrolyses oil shale with oil refinery residue (bitumen). Some tests involve co-pyrolysis of oil shale with lignite and cellulose wastes. Depending on reaction conditions, the co-pyrolysis may lead to higher conversion ratios and thus lower production costs, and in some cases solves the problem of utilization of certain wastes.

Chapter 6

Environmental Issues of Petroleum

The environmental impact of petroleum is often negative because it is toxic to almost all forms of life. The possibility of climate change exists. Petroleum, commonly referred to as oil, is closely linked to virtually all aspects of present society, especially for transportation and heating for both homes and for commercial activities.

Toxicity

Crude oil is a mixture of many different kinds of organic compounds, many of which are highly toxic and cancer causing (carcinogenic). Oil is "acutely lethal" to fish, that is it kills fish quickly, at a concentration of 4000 parts per million (ppm) (0.4%). "It only takes one quart of motor oil to make 250,000 gallons of ocean water toxic to wildlife." This is the equivalent of a concentration of 1 ppm. Crude oil and petroleum distillates cause birth defects.

Benzene is present in both crude oil and gasoline and is known to cause leukemia in humans. The compound is also known to lower the white blood cell count in humans, which would leave people exposed to it more susceptible to infections. "Studies have linked benzene exposure in the mere parts per billion (ppb) range to terminal leukemia, Hodgkins lymphoma, and other blood and immune system diseases within 5-15 years of exposure."

Exhaust

When oil or petroleum distillates are burned, usually the combustion is not complete. This means that incompletely burned compounds are created in addition to just water and carbon dioxide. The other compounds are often toxic to life. Examples are carbon monoxide and methanol. Also, fine particulates of soot blacken humans' and other animals' lungs and cause heart problems or death. Soot is cancer causing (carcinogenic).

Acid Rain

High temperatures created by the combustion of petroleum cause nitrogen gas in the surrounding air to oxidize, creating nitrous oxides. Nitrous oxides, along with sulfur dioxide from the sulfur in the oil, combine with water in the atmosphere to create acid rain. Acid rain causes many problems such as dead trees and acidified lakes with dead fish. Coral reefs in the world's oceans are killed by acidic water caused by acid rain. Acid rain leads to increased corrosion of machinery and structures (large amounts of capital), and to the slow destruction of important archaeological structures such as the marble ruins in Rome and Greece.

Climate Change

Humans burning large amounts of petroleum create large amounts of CO_2 (carbon dioxide) gas that traps heat in the earth's atmosphere. Also certain organic compounds, such as methane released from petroleum drilling or from the petroleum itself, trap heat several times more efficiently than CO_2. Soot blocks the sun from reaching the earth and could cause cooling of the earth's atmosphere.

Oil Spills

Oil spills are the release of liquid petroleum hydrocarbons into the environment due to human activity, and are a form of pollution. The term often refers to marine oil spills, where oil is released into the ocean or coastal waters. Oil spills include releases of crude oil from tankers, offshore platforms, drilling rigs and wells, as well as spills of refined petroleum products (such as gasoline, diesel) and their by-products, as well as heavier fuels used by large ships such as bunker fuel, or the spill of any oily refuse or waste oil. Spills may take years or even decades to clean up, and their total environmental impacts are not completely understood.

Oil Shale

Winning of petroleum from oil shale creates pollution problems

Volatile Organic Compounds

Volatile organic compounds (VOCs) are gases or vapours emitted by various solids and liquids, many of which have short- and long-term adverse effects on human health and the environment. VOCs from petroleum are toxic and foul the air, and some like benzene are extremely toxic, carcinogenic and cause DNA damage. Benzene often makes up about 1% of crude oil and gasoline. Benzene is present in automobile exhaust.

Waste Oil

Waste oil is used oil containing breakdown products and impurities from use. Some examples of waste oil are used oils such as hydraulic oil, transmission oil, brake fluids, motor oil, crankcase oil, gear box oil and synthetic oil. Many of the same problems associated with natural petroleum exist with waste oil. When waste oil from vehicles drips out engines over streets and roads, the oil travels into the water table bringing with it such toxins as benzene. This poisons both soil and drinking water. Runoff from storms carries waste oil into rivers and oceans, poisoning them as well.

Mitigation

Conservation/phasing Out

- Creating laws to completely phase out the use of petroleum (Sweden's 15 year plan)
- Making use of petroleum more efficiently via better technology

Substitution of Other Energy Sources

- Using "cleaner" energy sources such as natural gas and biodiesel, especially in critical areas like cities where there are people.

Use of Biomass Instead of Petroleum

- It is said that anything that can be made from oil can be made from cellulose, that is fibrous plant material such as from hemp.
- Plastics can be created from cellulose instead of from oil.
- Lubricants like motor oil and grease can be made from plants and animal fat.

Safety Measures

- Decreasing the risk of spills
- False floors at gasoline stations to catch gasoline and oil drips from making it into the water table.

Waste Oil

Waste oil is defined as any petroleum-based or synthetic oil that, through use or handling, has become unsuitable for its original purpose due to the presence of impurities or loss of original properties.

Differentiating between "Waste Oil" and "Used Oil"

The U.S. EPA defines the term "used oil" using (approximately) the definition given above, while the term "waste oil" is used to mean used oil (or oil) that is inappropriately mixed with hazardous waste, or disposed of, rather than recycled (re-refined, burned for energy recovery, etc.). The general difference between the two terms is essentially whether the used oil is disposed of, or appropriately recycled.

Products used as Waste Oil

Some examples of types of products that after use, can be labeled as used oil are: hydraulic oil, transmission oil, brake fluids, motor oil, crankcase oil, gear box oil, synthetic oil, and grades 1, 2, 3 and 4 fuel oil.

Disposal of Waste Oil

Waste oil can be disposed of in different ways, including sending the used oil off-site (some facilities are permitted to handle the used oil such as your local garages and local waste disposal facilities), burning used oil as a fuel (some used oil is not regulated by burner standards, but others that are off-specification used oil can only be burned in either industrial furnaces, certain boilers, and permitted hazardous waste incinerators), and marketing the used oil (claims are made that the used oil is to be burned for energy recovery, it is then shipped to a used oil burner who burns the used oil in an approved industrial furnace or boiler).

Oils that are off-specification typically contain: Arsenic 5 ppm, Cadmium 2 ppm, Chromium 10 ppm, Lead 100 ppm, Flash point 100°F, minimum (i.e., fp must be greater than 100°F), Total Halogens >4,000 ppm

Storage and Handling of Waste Oil

For on-site burning of used oil, the oil must be stored in tanks or containers, above or underground. The containers must be in good condition with no leaks, the tanks/containers must be labeled with the words "used oil", and there must be a spill prevention plan (or a control and countermeasures plan).

Waste Oil Furnaces and Boilers

Waste oil furnace is a type of furnace used for heating purposes and is fuelled by used oil that is free of hazardous contaminants, as described by the EPA. Waste-oil-fuelled boilers can be used for various industrial purposes as well as heating.

Renewable Energy

Renewable energy is energy which comes from natural resources such as sunlight, wind, rain, tides, and geothermal heat, which are renewable (naturally replenished).

About 16% of global final energy consumption comes from renewables, with 10% coming from traditional biomass, which is mainly used for heating, and 3.4% from hydroelectricity. New renewables (small hydro, modern biomass, wind, solar, geothermal, and biofuels) accounted for another 2.8% and are growing very rapidly. The share of renewables in electricity generation is around 19%, with 16% of global electricity coming from hydroelectricity and 3% from new renewables.

Wind power is growing at the rate of 30% annually, with a worldwide installed capacity of 198 gigawatts (GW) in 2010, and is widely used in Europe, Asia, and the United States. At the end of 2010, cumulative global photovoltaic (PV) installations surpassed 40 GW and PV power stations are popular in Germany and Spain. Solar thermal power stations operate in the USA and Spain, and the largest of these is the 354 megawatt (MW) SEGS power plant in the Mojave Desert.

The world's largest geothermal power installation is the Geysers in California, with a rated capacity of 750 MW. Brazil has one of the largest renewable energy programs in the world, involving production of ethanol fuel from sugar cane, and ethanol now provides 18% of the country's automotive fuel. Ethanol fuel is also widely available in the USA.

While many renewable energy projects are large-scale, renewable technologies are also suited to rural and remote areas, where energy is often crucial in human development. As of 2011, small solar PV systems provide electricity to a few million households, and micro-hydro configured into mini-grids serves many more. Over 44 million households use biogas made in household-scale digesters for lighting and/or cooking, and more than 166 million households rely on a new generation of more-efficient biomass cookstoves. United Nations' Secretary-General Ban Ki-moon has said that renewable energy has the ability to lift the poorest nations to new levels of prosperity.

Climate change concerns, coupled with high oil prices, peak oil, and increasing government support, are driving increasing renewable energy legislation, incentives and commercialization. New government spending, regulation and policies helped the industry weather the global financial crisis better than many other sectors. According to a 2011 projection by the International Energy Agency, solar power generators may produce most of the world's electricity within 50 years, dramatically reducing the emissions of greenhouse gases that harm the environment.

Renewable energy is derived from natural processes that are replenished constantly. In its various forms, it derives directly from the sun, or from heat generated deep within the earth. Included in the definition is electricity and heat generated from solar, wind, ocean, hydropower, biomass, geothermal resources, and biofuels and hydrogen derived from renewable resources.

Renewable energy replaces conventional fuels in four distinct areas: power generation, hot water/space heating, transport fuels, and rural (off-grid) energy services:

- Power generation. Renewable energy provides 18 percent of total electricity generation worldwide. Renewable power generators are spread across many countries, and wind power alone already provides a significant share of electricity in some areas: for example, 14 percent in the U.S. state of Iowa, 40 percent in the northern German state of Schleswig-Holstein, and 20 percent in Denmark.

  Some countries get most of their power from renewables, including Iceland and Paraguay (100 percent), Norway (98 percent), Brazil (86 percent), Austria (62 percent), New Zealand (65 percent), and Sweden (54 percent).

- Heating. Solar hot water makes an important contribution to renewable heat in many countries, most notably in China, which now has 70 percent of the global total (180 GWth). Most of these systems are installed on multi-family apartment buildings and meet a portion of the hot water needs of an estimated 50–60 million households in China.

  Worldwide, total installed solar water heating systems meet a portion of the water heating needs of over 70 million households. The use of biomass for heating continues to grow as well. In Sweden, national use of biomass energy has surpassed that of oil. Direct geothermal for heating is also growing rapidly.

- Transport fuels. Renewable biofuels have contributed to a significant decline in oil consumption in the United States since 2006. The 93 billion liters of biofuels produced worldwide in 2009 displaced the equivalent of an estimated 68 billion liters of gasoline, equal to about 5 percent of world gasoline production.

Mainstream Forms of Renewable Energy

Wind Power

Airflows can be used to run wind turbines. Modern wind turbines range from around 600 kW to 5 MW of rated power, although turbines with rated output of 1.5–3 MW have become the most common for commercial use; the power output of a turbine is a function of the cube of the wind speed, so as wind speed increases, power output increases dramatically. Areas where winds are stronger and more constant, such as offshore and high altitude sites, are preferred locations for wind farms. Typical capacity factors are 20-40%, with values at the upper end of the range in particularly favourable sites.

Globally, the long-term technical potential of wind energy is believed to be five times total current global energy production, or 40 times current electricity demand. This could require wind turbines to be installed over large areas, particularly in areas of higher wind resources. Offshore resources experience mean wind speeds of ~90% greater than that of land, so offshore resources could contribute substantially more energy.

Hydropower

Energy in water can be harnessed and used. Since water is about 800 times denser than air, even a slow flowing stream of water, or moderate sea swell, can yield considerable amounts of energy. There are many forms of water energy:

- Hydroelectric energy is a term usually reserved for large-scale hydroelectric dams. Examples are the Grand Coulee Dam in Washington State and the Akosombo Dam in Ghana.
- Micro hydro systems are hydroelectric power installations that typically produce up to 100 kW of power. They are often used in water rich areas as a remote-area power supply (RAPS).
- Run-of-the-river hydroelectricity systems derive kinetic energy from rivers and oceans without using a dam.

Solar Energy

Solar energy is the energy derived from the sun through the form of solar radiation. Solar powered electrical generation relies on photovoltaics and heat engines. A partial list of other solar applications includes space heating and cooling through solar architecture, daylighting, solar hot water, solar cooking, and high temperature process heat for industrial purposes.

Solar technologies are broadly characterized as either passive solar or active solar depending on the way they capture, convert and distribute solar energy. Active solar techniques include the use of photovoltaic panels and solar thermal collectors to harness the energy. Passive solar techniques include orienting a building to the Sun, selecting materials with favourable thermal mass or light dispersing properties, and designing spaces that naturally circulate air.

Biomass

Biomass (plant material) is a renewable energy source because the energy it contains comes from the sun. Through the process of photosynthesis, plants capture the sun's energy. When the plants are burnt, they release the sun's energy they contain. In this way, biomass functions as a sort of natural battery for storing solar energy. As long as biomass is produced sustainably, with only as much used as is grown, the battery will last indefinitely.

In general there are two main approaches to using plants for energy production: growing plants specifically for energy use, and using the residues from plants that are used for other things. The best approaches vary from region to region according to climate, soils and geography.

Biofuel

Biofuels include a wide range of fuels which are derived from biomass. The term covers solid biomass, liquid fuels and various biogases. Liquid biofuels include bioalcohols, such as bioethanol, and oils, such as biodiesel. Gaseous biofuels include biogas, landfill gas and synthetic gas.

Bioethanol is an alcohol made by fermenting the sugar components of plant materials and it is made mostly from sugar and starch crops. With advanced technology being developed, cellulosic biomass, such as trees and grasses, are also used as feedstocks for ethanol production. Ethanol can be used as a fuel for vehicles in its pure form, but it is usually used as a gasoline additive to increase octane and improve vehicle emissions. Bioethanol is widely used in the USA and in Brazil.

Biodiesel is made from vegetable oils, animal fats or recycled greases. Biodiesel can be used as a fuel for vehicles in its pure form, but it is usually used as a diesel additive to reduce levels of particulates, carbon monoxide, and hydrocarbons from diesel-powered vehicles. Biodiesel is produced from oils or fats using transesterification and is the most common biofuel in Europe.

Biofuels provided 2.7% of the world's transport fuel in 2010.

Geothermal Energy

Geothermal energy is energy obtained by trapping the heat of the earth itself, both from kilometres deep into the Earth's crust in volcanically active locations of the globe or from shallow depths, as in geothermal heat pumps in most locations of the planet. It is expensive to build a power station but operating costs are low resulting in low energy costs for suitable sites. Ultimately, this energy derives from heat in the Earth's core.

Three types of power plants are used to generate power from geothermal energy: dry steam, flash, and binary. Dry steam plants take steam out of fractures in the ground and use it to directly drive a turbine that spins a generator. Flash plants take hot water, usually at temperatures over 200 °C, out of the ground, and allows it to boil as it rises to the surface then separates the steam phase in steam/water separators and then runs the steam through a turbine. In binary plants, the hot water flows through heat exchangers, boiling an organic fluid that spins the turbine. The condensed steam and remaining geothermal fluid from all three types of plants are injected back into the hot rock to pick up more heat.

The geothermal energy from the core of the Earth is closer to the surface in some areas than in others. Where hot underground steam or water can be tapped and brought to the surface it may be used to generate electricity. Such geothermal power sources exist in certain geologically unstable parts of the world such as Chile, Iceland, New Zealand, United States, the Philippines and Italy. The two most prominent areas for this in the United States are in the Yellowstone basin and in northern California. Iceland produced 170 MW geothermal power and heated 86% of all houses in the year 2000 through geothermal energy. Some 8000 MW of capacity is operational in total.

Renewable Energy Commercialization

Growth of Renewables

During the five-years from the end of 2004 through 2009, worldwide renewable energy capacity grew at rates of 10–60 percent annually for many technologies. For wind power and many other renewable technologies, growth accelerated in 2009 relative to the previous four years. More wind power capacity was added during 2009 than any other renewable technology. However, grid-connected PV increased the fastest of all renewables technologies, with a 60-percent annual average growth rate for the five-year period.

Table: *Selected renewable energy indicators*

Selected global indicators	***2008***	***2009***	***2010***
Investment in new renewable capacity (annual)	130	160	211 billion USD
Renewables power capacity (existing)	1,140	1,230	1,320 GWe
Hydropower capacity (existing)	950	980	1,010 GWe
Wind power capacity (existing)	121	159	198 GWe
Solar PV capacity (grid-connected)	16	23	40 GWe
Solar hot water capacity (existing)	130	160	185 GWth
Ethanol production (annual)	67	76	86 billion liters
Countries with policy targetsfor renewable energy use	79	89	98

Scientists have advanced a plan to power 100% of the world's energy with wind, hydroelectric, and solar power by the year 2030.

According to a 2011 projection by the International Energy Agency, solar power generators may produce most of the world's electricity within 50 years, dramatically reducing the emissions of greenhouse gases that harm the environment. Cedric Philibert, senior analyst in the renewable energy division at the IEA said: "Photovoltaic and solar-thermal plants may meet most of the world's demand for electricity by 2060 — and half of all energy needs — with wind, hydropower and biomass plants supplying much of the remaining generation". "Photovoltaic and concentrated solar power together can become the major source of electricity," Philibert said.

Economic Trends

All forms of energy are expensive, but as time progresses, renewable energy generally gets cheaper, while fossil fuels generally get more expensive. Al Gore has explained that renewable energy technologies are declining in price for three main reasons:

First, once the renewable infrastructure is built, the fuel is free. Unlike carbon-based fuels, the wind and the sun and the earth itself provide fuel that is free, in amounts that are effectively limitless.

Second, while fossil fuel technologies are more mature, renewable energy technologies are being rapidly improved. So innovation and ingenuity give us the ability to constantly increase the efficiency of renewable energy and continually reduce its cost.

Third, once the world makes a clear commitment to shifting toward renewable energy, the volume of production will itself sharply reduce the cost of each windmill and each solar panel, while adding yet more incentives for additional research and development to further speed up the innovation process.

Wind Power Market

Global wind power installations increased by 35,800 MW in 2010, bringing total installed capacity up to 194,400 MW, a 22.5% increase on the 158,700 MW installed at the end of 2009. For the first time more than half of all new wind power was added outside of the traditional markets of Europe and North America, mainly driven, by the continuing boom in China which accounted for nearly half of all of the installations at 16,500 MW.

China now has 42,300 MW of wind power installed. Several countries have achieved relatively high levels of wind power penetration, such as 21% of stationary electricity production in Denmark, 18% in Portugal, 16% in Spain, 14% in Ireland and 9% in Germany in 2010. As of 2011, 83 countries around the world are using wind power on a commercial basis.

Table: *Top 10 wind power countries*

Country	***Total capacityend 2009 (MW)***	***Total capacityJune 2010 (MW)***
United States	35,159	36,300
China	26,010	33,800
Germany	25,777	26,400
Spain	19,149	19,500
India	10, 925	12,100
Italy	4,850	5,300
France	4,521	5,000
United Kingdom	4,092	4,600
Portugal	3,535	3,800
Denmark	3,497	3,700
Rest of world	21,698	24,500
Total	159,213	175,000

As of November 2010, the Roscoe Wind Farm (781 MW) is the world's largest wind farm. As of September 2010, the Thanet Offshore Wind Project in United Kingdom is the largest offshore wind farm in the world at 300 MW, followed by Horns Rev II (209 MW) in Denmark. The United Kingdom is the world's leading generator of offshore wind power, followed by Denmark.

There are many large wind farms under construction and these include BARD Offshore 1 (400 MW), Greater Gabbard wind farm (500 MW), Lincs Wind Farm (270 MW), London Array (1000 MW), Lower Snake River Wind Project (343 MW), Shepherds Flat Wind Farm (845 MW), Sheringham Shoal (317 MW), and the Walney Wind Farm (367 MW).

New Generation of Solar Thermal Plants

Large solar thermal power stations include the 354 megawatt (MW) Solar Energy Generating Systems power plant in the USA, Solnova Solar Power Station (Spain, 150 MW), Andasol solar power station (Spain, 100 MW), Nevada Solar One (USA, 64 MW), PS20 solar power tower (Spain, 20 MW), and the PS10 solar power tower (Spain, 11 MW).

The Ivanpah Solar Power Facility is a 392 megawatt (MW) solar power facility which is under construction in south-eastern California. The Solana Generating Station is a 280 MW solar power plant which is under construction near Gila Bend, Arizona, about 70 miles (110 km) southwest of Phoenix. The Crescent Dunes Solar Energy Project is a 110 megawatt (MW) solar thermal power project currently under construction near Tonopah, about 190 miles (310 km) northwest of Las Vegas.

The solar thermal power industry is growing rapidly with 1.2 GW under construction as of April 2009 and another 13.9 GW announced globally through 2014. Spain is the epicenter of solar thermal power development with 22 projects for 1,037 MW under construction, all of which are projected to come online by the end of 2010. In the United States, 5,600 MW of solar thermal power projects have been announced. In developing countries, three World Bank projects for integrated solar thermal/combined-cycle gas-turbine power plants in Egypt, Mexico, and Morocco have been approved.

Photovoltaic Market

Solar photovoltaic cells convert sunlight into electricity and photovoltaic production has been increasing by an average of more than 20 percent each year since 2002, making it a fast-growing energy technology. At the end of 2010, cumulative global photovoltaic (PV) installations surpassed 40 GW and PV power stations are popular in Germany and Spain.

Many solar photovoltaic power stations have been built, mainly in Europe. As of October 2011, the largest photovoltaic (PV) power plants in the world are the Sarnia Photovoltaic Power Plant (Canada, 97 MW),

Montalto di Castro Photovoltaic Power Station (Italy, 84.2 MW), Finsterwalde Solar Park (Germany, 80.7 MW), Lieberose Photovoltaic Park (Germany, 71.8 MW), Rovigo Photovoltaic Power Plant (Italy, 70 MW), Olmedilla Photovoltaic Park (Spain, 60 MW), and the Strasskirchen Solar Park (Germany, 54 MW).

There are also many large plants under construction. The Desert Sunlight Project is a 550 MW solar power plant under construction in Riverside County, California, that will use thin-film solar photovoltaic modules made by First Solar. The Blythe Solar Power Project is a 500 MW photovoltaic station under construction in Riverside County, California. The Agua Caliente Solar Project is a 290 megawatt photovoltaic solar generating facility being built in Yuma County, Arizona. The California Valley Solar Ranch (CVSR) is a 250 megawatt (MW) solar photovoltaic power plant, which is being built by SunPower in the Carrizo Plain, northeast of California Valley. The Mesquite Solar project is a photovoltaic solar power plant being built in Arlington, Maricopa County, Arizona, owned by Sempra Generation. Phase 1 will have a nameplate capacity of 150 megawatts.

Many of these plants are integrated with agriculture and some use innovative tracking systems that follow the sun's daily path across the sky to generate more electricity than conventional fixed-mounted systems. There are no fuel costs or emissions during operation of the power stations. However, when it comes to renewable energy systems and PV, it is not just large systems that matter. Building-integrated photovoltaics or "onsite" PV systems use existing land and structures and generate power close to where it is consumed.

Biofuels for Transportation

Biofuels provided 2.7% of the world's transport fuel in 2010. Mandates for blending biofuels exist in 31 countries at the national level and in 29 states/provinces. According to the International Energy Agency, biofuels have the potential to meet more than a quarter of world demand for transportation fuels by 2050.

Since the 1970s, Brazil has had an ethanol fuel program which has allowed the country to become the world's second largest producer of ethanol (after the United States) and the world's largest exporter. Brazil's ethanol fuel program uses modern equipment and cheap sugar cane as feedstock, and the residual cane-waste (bagasse) is used to produce heat and power. There are no longer light vehicles in Brazil running on pure gasoline. By the end of 2008 there were 35,000 filling stations throughout Brazil with at least one ethanol pump.

Nearly all the gasoline sold in the United States today is mixed with 10 percent ethanol, a mix known as E10, and motor vehicle manufacturers already produce vehicles designed to run on much higher ethanol blends.

Ford, DaimlerChrysler, and GM are among the automobile companies that sell "flexible-fuel" cars, trucks, and minivans that can use gasoline and ethanol blends ranging from pure gasoline up to 85% ethanol (E85). By mid-2006, there were approximately 6 million E85-compatible vehicles on U.S. roads.

The challenge is to expand the market for biofuels beyond the farm states where they have been most popular to date. Flex-fuel vehicles are assisting in this transition because they allow drivers to choose different fuels based on price and availability. The *Energy Policy Act of 2005*, which calls for 7.5 billion US gallons (28,000,000 m^3) of biofuels to be used annually by 2012, will also help to expand the market.

Geothermal Energy Commercialization

The International Geothermal Association (IGA) has reported that 10,715 megawatts (MW) of geothermal power in 24 countries is online, which is expected to generate 67,246 GWh of electricity in 2010. This represents a 20% increase in geothermal power online capacity since 2005. IGA projects this will grow to 18,500 MW by 2015, due to the large number of projects presently under consideration, often in areas previously assumed to have little exploitable resource.

In 2010, the United States led the world in geothermal electricity production with 3,086 MW of installed capacity from 77 power plants; the largest group of geothermal power plants in the world is located at The Geysers, a geothermal field in California. The Philippines follows the US as the second highest producer of geothermal power in the world, with 1,904 MW of capacity online; geothermal power makes up approximately 18% of the country's electricity generation.

Geothermal (ground source) heat pumps represented an estimated 30 GWth of installed capacity at the end of 2008, with other direct uses of geothermal heat (i.e., for space heating, agricultural drying and other uses) reaching an estimated 15 GWth. As of 2008, at least 76 countries use direct geothermal energy in some form.

Developing Country Markets

Renewable energy can be particularly suitable for developing countries. In rural and remote areas, transmission and distribution of energy generated from fossil fuels can be difficult and expensive. Producing renewable energy locally can offer a viable alternative.

As of 2011, small solar PV systems provide electricity to a few million households, and micro-hydro configured into mini-grids serves many more. Over 44 million households use biogas made in household-scale digesters for lighting and/or cooking, and more than 166 million households rely on a new generation of more-efficient biomass cookstoves.

Kenya is the world leader in the number of solar power systems installed per capita. More than 30,000 very small solar panels, each producing 12 to 30 watts, are sold in Kenya annually.

Renewable energy projects in many developing countries have demonstrated that renewable energy can directly contribute to poverty alleviation by providing the energy needed for creating businesses and employment. Renewable energy technologies can also make indirect contributions to alleviating poverty by providing energy for cooking, space heating, and lighting. Renewable energy can also contribute to education, by providing electricity to schools.

Industry and Policy Trends

U.S. President Barack Obama's American Recovery and Reinvestment Act of 2009 includes more than $70 billion in direct spending and tax credits for clean energy and associated transportation programs. Clean Edge suggests that the commercialization of clean energy will help countries around the world pull out of the current economic malaise. Leading renewable energy companies include First Solar, Gamesa, GE Energy, Q-Cells, Sharp Solar, Siemens, SunOpta, Suntech, and Vestas.

The International Renewable Energy Agency (IRENA) is an intergovernmental organization for promoting the adoption of renewable energy worldwide. It aims to provide concrete policy advice and facilitate capacity building and technology transfer. IRENA was formed on January 26, 2009, by 75 countries signing the charter of IRENA. As of March 2010, IRENA has 143 member states who all are considered as founding members, of which 14 have also ratified the statute.

As of 2011, 119 countries have some form of national renewable energy policy target or renewable support policy. National targets now exist in at least 98 countries. There is also a wide range of policies at state/provincial and local levels.

New and Emerging Renewable Energy Technologies

New and emerging renewable energy technologies are still under development and include cellulosic ethanol, hot-dry-rock geothermal power, and ocean energy.

These technologies are not yet widely demonstrated or have limited commercialization. Many are on the horizon and may have potential comparable to other renewable energy technologies, but still depend on attracting sufficient attention and research, development and demonstration (RD&D) funding.

Cellulosic Ethanol

Companies such as Iogen, Broin, and Abengoa are building refineries that can process biomass and turn it into ethanol, while companies such as Diversa, Novozymes, and Dyadic are producing enzymes which could enable a cellulosic ethanol future. The shift from food crop feedstocks to waste residues and native grasses offers significant opportunities for a range of players, from farmers to biotechnology firms, and from project developers to investors.

Table: *Selected Commercial Cellulosic Ethanol Plants in the U.S. (Operational or under construction)*

Company	***Location***	***Feedstock***
Abengoa Bioenergy	Hugoton, KS	Wheat straw
BlueFire Ethanol	Irvine, CA	Multiple sources
Gulf Coast Energy	Mossy Head, FL	Wood waste
Mascoma	Lansing, MI	Wood
POET LLC	Emmetsburg, IA	Corn cobs
SunOpta	Little Falls, MN	Wood chips
Xethanol	Auburndale, FL	Citrus peels

Ocean Energy

Systems to harvest utility-scale electrical power from ocean waves have recently been gaining momentum as a viable technology. The potential for this technology is considered promising, especially on west-facing coasts with latitudes between 40 and 60 degrees:

In the United Kingdom, for example, the Carbon Trust recently estimated the extent of the economically viable offshore resource at 55 TWh per year, about 14% of current national demand. Across Europe, the technologically achievable resource has been estimated to be at least 280 TWh per year. In 2003, the U.S. Electric Power Research Institute (EPRI) estimated the viable resource in the United States at 255 TWh per year (6% of demand).

Funding for a wave farm in Scotland was announced in February, 2007 by the Scottish Government, at a cost of over 4 million pounds, as part of a UK£13 million funding packages for ocean power in Scotland. The farm will be the world's largest with a capacity of 3MW generated by four Pelamis machines.

The world's first commercial tidal power station was installed in 2007 in the narrows of Strangford Lough in Ireland. The 1.2 megawatt underwater tidal electricity generator, part of Northern Ireland's Environment & Renewable Energy Fund scheme, takes advantage of the fast tidal flow (up to 4 metres per second) in the lough. Although the generator is powerful enough to power a thousand homes, the turbine has minimal environmental impact, as it is almost entirely submerged, and the rotors pose no danger to wildlife as they turn quite slowly.

Ocean thermal energy conversion (OTEC) uses the temperature difference that exists between deep and shallow waters to run a heat engine.

Enhanced Geothermal Systems

Enhanced Geothermal Systems are a new type of geothermal power technologies that do not require natural convective hydrothermal resources. The vast majority of geothermal energy within drilling reach is in dry and non-porous rock. EGS technologies "enhance" and/or create geothermal resources in this "hot dry rock (HDR)" through hydraulic stimulation.

EGS/HDR technologies, like hydrothermal geothermal, are expected to be baseload resources which produce power 24 hours a day like a fossil plant. Distinct from hydrothermal, HDR/EGS may be feasible anywhere in the world, depending on the economic limits of drill depth. Good locations are over deep granite covered by a thick (3–5 km) layer of insulating sediments which slow heat loss.

There are HDR and EGS systems currently being developed and tested in France, Australia, Japan, Germany, the U.S. and Switzerland. The largest EGS project in the world is a 25 megawatt demonstration plant currently being developed in the Cooper Basin, Australia. The Cooper Basin has the potential to generate 5,000–10,000 MW.

Experimental Solar Power

Concentrated photovoltaics (CPV) systems employ sunlight concentrated onto photovoltaic surfaces for the purpose of electrical power production. Thermoelectric, or "thermovoltaic" devices convert a temperature difference between dissimilar materials into an electric current. Space-based solar power is a theoretical design for the collection of solar power in space, for use on Earth.

Artificial Photosynthesis

Artificial photosynthesis uses techniques include nanotechnology to store solar electromagnetic energy in chemical bonds by splitting water to produce hydrogen and then using carbon dioxide to make methanol.

Renewable Energy Debate

Renewable electricity production, from sources such as wind power and solar power, is sometimes criticized for being variable or intermittent. However, the International Energy Agency has stated that deployment of renewable technologies usually increases the diversity of electricity sources and, through local generation, contributes to the flexibility of the system and its resistance to central shocks.

There have been "not in my back yard" (NIMBY) concerns relating to the visual and other impacts of some wind farms, with local residents sometimes fighting or blocking construction. In the USA, the Massachusetts Cape Wind project was delayed for years partly because of aesthetic concerns. However, residents in other areas have been more positive and there are many examples of community wind farm developments. According to a town councilor, the overwhelming majority of locals believe that the Ardrossan Wind Farm in Scotland has enhanced the area.

The market for renewable energy technologies has continued to grow. Climate change concerns, coupled with high oil prices, peak oil, and increasing government support, are driving increasing renewable energy legislation, incentives and commercialization. New government spending, regulation and policies helped the industry weather the 2009 economic crisis better than many other sectors.

Hubbert Peak Theory

The Hubbert peak theory posits that for any given geographical area, from an individual oil-producing region to the planet as a whole, the rate of petroleum production tends to follow a bell-shaped curve. It is one of the primary theories on peak oil.

Choosing a particular curve determines a point of maximum production based on discovery rates, production rates and cumulative production. Early in the curve (pre-peak), the production rate increases because of the discovery rate and the addition of infrastructure. Late in the curve (post-peak), production declines because of resource depletion.

The Hubbert peak theory is based on the observation that the amount of oil under the ground in any region is finite, therefore the rate of discovery which initially increases quickly must reach a maximum and decline. In the US, oil extraction followed the discovery curve after a time lag of 32 to 35 years. The theory is named after American geophysicist M. King Hubbert, who created a method of modelling the production curve given an assumed ultimate recovery volume.

Hubbert's Peak

"Hubbert's peak" can refer to the peaking of production of a particular area, which has now been observed for many fields and regions.

Hubbert's Peak was achieved in the continental US in the early 1970s. Oil production peaked at 10,200,000 barrels per day (1,620,000 m^3/d). Since then, it has been in a gradual decline.

Peak oil as a proper noun, or "Hubbert's peak" applied more generally, refers to a singular event in history: the peak of the entire planet's oil production. After Peak Oil, according to the Hubbert Peak Theory, the rate of oil production on Earth would enter a terminal decline. On the basis of his theory, in a paper he presented to the American Petroleum Institute in 1956, Hubbert correctly predicted that production of oil from conventional sources would peak in the continental United States around 1965-1970. Hubbert further predicted a worldwide peak at "about half a century" from publication and approximately 12 gigabarrels (GB) a year in magnitude. In a 1976 TV interview Hubbert added that the actions of OPEC might flatten the global production curve but this would only delay the peak for perhaps 10 years.

Hubbert's Theory

Hubbert Curve

In 1956, Hubbert proposed that fossil fuel production in a given region over time would follow a roughly bell-shaped curve without giving a precise formula; he later used the Hubbert curve, the derivative of the logistic curve, for estimating future production using past observed discoveries.

Hubbert assumed that after fossil fuel reserves (oil reserves, coal reserves, and natural gas reserves) are discovered, production at first increases approximately exponentially, as more extraction commences and more efficient facilities are installed. At some point, a peak output is reached, and production begins declining until it approximates an exponential decline.

The Hubbert curve satisfies these constraints. Furthermore, it is roughly symmetrical, with the peak of production reached when about half of the fossil fuel that will ultimately be produced has been produced. It also has a single peak.

Given past oil discovery and production data, a Hubbert curve that attempts to approximate past discovery data may be constructed and used to provide estimates for future production. In particular, the date of peak oil production or the total amount of oil ultimately produced can be estimated that way. Cavallo defines the Hubbert curve used to predict the U.S. peak as the derivative of:

$$Q(t) = \frac{Q_{max}}{1 + ae^{-bt}}$$

where Q_{max} is the total resource available (ultimate recovery of crude oil), $Q(t)$ the cumulative production, and a and b are constants. The year of maximum annual production (peak) is:

$$t_{max} = \frac{1}{b}\ln(a).$$

Use of Multiple Curves

The sum of multiple Hubbert curves, a technique not developed by Hubbert himself, may be used in order to model more complicated real life scenarios.

Definition of Reserves

Almost all of Hubbert peaks must be put in the context of high ore grade. Except for fissionable materials, any resource, including oil, is theoretically recoverable from the environment with the right technology. In contrast, Hubbert was concerned with "easy" oil, "easy" metals, and so forth that could be recovered without greatly advanced mining efforts and how to time the necessity of such resource acquisition advancements or substitutions by knowing an "easy" resource's probable peak.

Also, as reserves become more difficult to extract there is the possibility that mining or alternatives are too expensive for developing countries.

For heavy crude or deep water drilling attempts, such as Noxal oil field or tar sands or oil shale, the price of the oil extracted will have to include the extra effort required to mine these resources. According to the U.S. Bureau of Ocean Energy Management, Regulation and Enforcement(formerly, the Minerals Management Service), areas such as the Outer Continental Shelf may also incur higher costs due to environmental concerns. So not all oil reserves are equal, and the more difficult reserves are predicted by Hubbert as being typical of the post-peak side of the Hubbert curve.

Reliability

Hubbert, in his 1956 paper, presented two scenarios for US conventional oil production (crude oil + condensate):

- most likely estimate: a logistic curve with a logistic growth rate equal to 6%, an ultimate resource equal to 150 Giga-barrels (Gb) and a peak in 1965.
- upper-bound estimate: a logistic curve with a logistic growth rate equal to 6% and ultimate resource equal to 200 Giga-barrels and a peak in 1970.

Hubbert's upper-bound estimate, which he regarded as optimistic, accurately predicted that US oil production would peak in 1970. Forty years later, the upper-bound estimate has also proven to be very accurate in terms of cumulative production, less so in terms of annual production. For 2005, the upper-bound Hubbert model predicts 178.2 Gb cumulative and 1.17 Gb current production; actual US production was 176.4 Gb cumulative crude oil + condensate (1% lower than the upper bound estimate), with annual production of 1.55 Gb (32% higher than the upper bound estimate).

A post-hoc analysis of peaked oil wells, fields, regions and nations found that Hubbert's model was the "most widely useful"(providing the best fit to the data), though many areas studied had a sharper "peak" than predicted.

Economics

Energy Return on Energy Investment

When oil production first began in the mid-nineteenth century, the largest oil fields recovered fifty barrels of oil for every barrel used in the extraction, transportation and refining. This ratio is often referred to as the Energy Return on Energy Investment (EROI or EROEI). Currently, between one and five barrels of oil are recovered for each barrel-equivalent of energy used in the recovery process. As the EROEI drops to one, or equivalently the Net energy gain falls to zero, the oil production is no longer a net energy source. This happens long before the resource is physically exhausted.

Note that it is important to understand the distinction between a barrel of oil, which is a measure of oil, and a barrel of oil equivalent (BOE), which is a measure of energy. Many sources of energy, such as fission, solar, wind, and coal, are not subject to the same near-term supply restrictions that oil is. Accordingly, even an oil source with an EROEI of 0.5 can be usefully exploited if the energy required to produce that oil comes from a cheap and plentiful energy source. Availability of cheap, but hard to transport, natural gas in some oil fields has led to using natural gas to fuel enhanced oil recovery. Similarly, natural gas in huge amounts is used to power most Athabasca Tar Sands plants. Cheap natural gas has also led to Ethanol fuel produced with a net EROEI of less than 1, although figures in this area are controversial because methods to measure EROEI are in debate.

Growth-based Economic Models

Insofar as economic growth is driven by oil consumption growth, post-peak societies must adapt. Hubbert believed: Some economists describe the problem as uneconomic growth or a false economy. At the political right, Fred Ikle has warned about "conservatives addicted to the Utopia of Perpetual Growth". Brief oil interruptions in 1973 and 1979 markedly slowed - but did not stop - the growth of world GDP. Between 1950 and 1984, as the Green Revolution transformed agriculture around the globe, world grain production increased by 250%. The energy for the Green Revolution was provided by fossil fuels in the form of fertilizers (natural gas), pesticides (oil), and hydrocarbon fuelled irrigation.

David Pimentel, professor of ecology and agriculture at Cornell University, and Mario Giampietro, senior researcher at the National Research Institute on Food and Nutrition (INRAN), place in their study *Food, Land, Population and the U.S. Economy* the maximum U.S. population for a sustainable economy at 200 million. To achieve a sustainable economy world population will have to be reduced by two-thirds, says the study. Without population reduction, this study predicts an agricultural crisis beginning in 2020, becoming critical c. 2050. The peaking of global oil along with the decline in regional natural gas production may precipitate this agricultural crisis sooner than generally expected. Dale Allen Pfeiffer claims that coming decades could see spiraling food prices without relief and massive starvation on a global level such as never experienced before.

Hubbert Peaks

Although Hubbert peak theory receives most attention in relation to peak oil production, it has also been applied to other natural resources.

Natural Gas

Doug Reynolds predicted in 2005 that the North American peak would occur in 2007. Bentley (p. 189) predicted a world "decline in conventional gas production from about 2020".

Even if new extraction techniques yield additional sources of natural gas, like coalbed methane, the energy returned on energy invested will be much lower than traditional gas sources, which inevitably leads to higher costs to consumers of natural gas.

Coal

Although observers believe that peak coal is significantly further out than peak oil, Hubbert studied the specific example of anthracite in the USA, a high grade coal, whose production peaked in the 1920s. Hubbert found that Anthracite matches a curve closely. Pennsylvania's coal production also matches Hubbert's curve closely, but this does not mean that coal in Pennsylvania is exhausted—far from it. If production in Pennsylvania returned at its all time high, there are reserves for 190 years. Hubbert had recoverable coal reserves worldwide at 2500×10^9 metric tons and peaking around 2150 (depending on usage).

More recent estimates suggest an earlier peak. *Coal: Resources and Future Production* (PDF 630KB), published on April 5, 2007 by the Energy Watch Group (EWG), which reports to the German Parliament, found that global coal production could peak in as few as 15 years. Reporting on this Richard Heinberg also notes that the date of peak annual energetic extraction from coal will likely come earlier than the date of peak in quantity of coal (tons per year) extracted as the most energy-dense types of coal have been mined most extensively. A second study, *The Future of Coal* by B. Kavalov and S. D. Peteves of the Institute for Energy (IFE), prepared for European Commission Joint Research Centre, reaches similar conclusions and states that ""'coal might not be so abundant, widely available and reliable as an energy source in the future".

Work by David Rutledge of Caltech predicts that the total of world coal production will amount to only about 450 gigatonnes. This implies that coal is running out faster than usually assumed.

Finally, insofar as global peak oil and peak in natural gas are expected anywhere from imminently to within decades at most, any increase in coal production (mining) per annum to compensate for declines in oil or natural gas production, would necessarily translate to an earlier date of peak as compared with peak coal under a scenario in which annual production remains constant.

Fissionable Materials

In a paper in 1956, after a review of US fissionable reserves, Hubbert notes of nuclear power:

> *"There is promise, however, provided mankind can solve its international problems and not destroy itself with nuclear weapons, and provided world population (which is now expanding at such a rate as to double in less than a century) can somehow be brought under control, that we may at last have found an energy supply adequate for our needs for at least the next few centuries of the "foreseeable future.""*

Technologies such as the thorium fuel cycle, reprocessing and fast breeders can, in theory, considerably extend the life of uranium reserves. Roscoe Bartlett claims

> *"Our current throwaway nuclear cycle uses up the world reserve of low-cost uranium in about 20 years."*

Caltech physics professor David Goodstein has stated that

> *"... you would have to build 10,000 of the largest power plants that are feasible by engineering standards in order to replace the 10 terawatts of fossil fuel we're burning today... that's a staggering amount and if you did that, the known reserves of uranium would last for 10 to 20 years at that burn rate.*
>
> *So, it's at best a bridging technology... You can use the rest of the uranium to breed plutonium 239 then we'd have at least 100 times as much fuel to use. But that means you're making plutonium, which is an extremely dangerous thing to do in the dangerous world that we live in."*

Helium

Almost all helium on Earth is a result of radioactive decay of uranium and thorium. Helium is extracted by fractional distillation from natural gas, which contains up to 7% helium. The world's largest helium-rich natural gas fields are found in the United States, especially in the Hugoton and nearby gas fields in Kansas, Oklahoma, and Texas. The extracted helium is stored underground in the National Helium Reserve near Amarillo, Texas, the self-proclaimed "Helium Capital of the World". Helium production is expected to decline along with natural gas production in these areas.

Helium is the second-lightest chemical element in the Universe, causing it to rise to the upper layers of Earth's atmosphere. Helium atoms are so light that the Earth's gravity field is simply not strong enough to trap helium in the atmosphere and it dissipates slowly into space and is lost forever.

Transition Metals

Hubbert applied his theory to "rock containing an abnormally high concentration of a given metal" and reasoned that the peak production for metals such as copper, tin, lead, zinc and others would occur in the time frame of decades and iron in the time frame of two centuries like coal. The price of copper rose 500% between 2003 and 2007 was by some attributed to peak copper. Copper prices later fell, along with many other commodities and stock prices, as demand shrank from fear of a global recession. Lithium availability is a concern for a fleet of Li-ion battery using cars but a paper published in 1996 estimated that world reserves are adequate for at least 50 years. A similar prediction for platinum use in fuel cells notes that the metal could be easily recycled.

Precious Metals

The possibility of Peak Gold has emerged recently [5]. Aaron Regent president of the Canadian gold giant Barrik Gold said that global output has been falling by roughly one million ounces a year since the start of the decade. The total global mine supply has dropped by 10pc as ore quality erodes, implying that the roaring bull market of the last eight years may have further to run. "There is a strong case to be made that we are already at 'peak gold'," he told The Daily Telegraph at the RBC's annual gold conference in London. "Production peaked around 2000 and it has been in decline ever since, and we forecast that decline to continue. It is increasingly difficult to find ore," he said. Ore grades have fallen from around 12 grams per tonne in 1950 to nearer 3 grams in the US, Canada, and Australia. South Africa's output has halved since peaking in 1970. Output fell a further 14 percent in South Africa in 2008 as companies were forced to dig ever deeper - at greater cost - to replace depleted reserves.

Phosphorus

Phosphorus supplies are essential to farming and depletion of reserves is estimated at somewhere from 60 to 130 years. According to a 2008 study, the total reserves of phosphorus are estimated to be approximately 3,200 MT, with a peak production at 28 MT/year in 2034.

Individual countries supplies vary widely; without a recycling initiative America's supply is estimated around 30 years. Phosphorus supplies affect agricultural output which in turn limits alternative fuels such as biodiesel and ethanol. Its increasing price and scarcity (global price of rock phosphate rose 8-fold in the 2 years to mid 2008) could change global agricultural patterns. Lands, perceived as marginal because of remoteness, but with very high phosphorus content, such as the Gran Chaco may get more agricultural development, while other farming areas, where nutrients are a constraint, may drop below the line of profitability.

Peak Water

Hubbert's original analysis did not apply to renewable resources. However, over-exploitation often results in a Hubbert peak nonetheless. A modified Hubbert curve applies to any resource that can be harvested faster than it can be replaced.

For example, a reserve such as the Ogallala Aquifer can be mined at a rate that far exceeds replenishment. This turns much of the world's underground water and lakes into finite resources with peak usage debates similar to oil. These debates usually centre around agriculture and suburban water usage but generation of electricity from nuclear energy or coal and tar sands mining mentioned above is also water resource intensive. The term fossil water is sometimes used to describe aquifers whose water is not being recharged.

Renewable Resources

- Fisheries: At least one researcher has attempted to perform Hubbert linearization (Hubbert curve) on the whaling industry, as well as charting the transparently dependent price of caviar on sturgeon depletion. Another example is the cod of the North Sea. The comparison of the cases of fisheries and of mineral extraction tells us that the human pressure on the environment is causing a wide range of resources to go through a depletion cycle which follows a Hubbert curve.

Criticisms

Economist Michael Lynch argues that the theory behind the Hubbert curve is too simplistic and relies on an overly Malthusian point of view. Lynch claims that Campbell's predictions for world oil production are strongly biased towards underestimates, and that Campbell has repeatedly pushed back the date.

Leonardo Maugeri, vice president of the Italian energy company Eni, argues that nearly all of peak estimates do not take into account unconventional oil even though the availability of these resources is significant and the costs of extraction and processing, while still very high, are falling because of improved technology. He also notes that the recovery rate from existing world oil fields has increased from about 22% in 1980 to 35% today because of new technology and predicts this trend will continue. The ratio between proven oil reserves and current production has constantly improved, passing from 20 years in 1948 to 35 years in 1972 and reaching about 40 years in 2003. These improvements occurred even with low investment in new exploration and upgrading technology because of the low oil prices during the last 20 years. However, Maugeri feels that encouraging more exploration will require relatively high oil prices.

Edward Luttwak, an economist and historian, claims that unrest in countries such as Russia, Iran and Iraq has led to a massive underestimate of oil reserves. The Association for the Study of Peak Oil and Gas, or ASPO responds by claiming neither Russia nor Iran are troubled by unrest currently, but Iraq is.

Cambridge Energy Research Associates authored a report that is critical of Hubbert-influenced predictions:

"Despite his valuable contribution, M. King Hubbert's methodology falls down because it does not consider likely resource growth, application of new technology, basic commercial factors, or the impact of geopolitics on production. His approach does not work in all cases-including on the United States itself-and cannot reliably model a global production outlook. Put more simply, the case for the imminent peak is flawed. As it is, production in 2005 in the Lower 48 in the United States was 66 percent higher than Hubbert projected."

CERA does not believe there will be an endless abundance of oil, but instead believes that global production will eventually follow an "undulating plateau" for one or more decades before declining slowly, and that production will reach 40 Mb/d by 2015.

Alfred J. Cavallo, while predicting a conventional oil supply shortage by no later than 2015, does not think Hubbert's peak is the correct theory to apply to world production.

Oil Burden

Oil burden is the volume of petroleum consumed multiplied by the average price and divided by nominal gross domestic product. This gives the proportion of the world economy devoted to buying oil. It is a concept developed by Veronique Riches-Flores of Société Générale. It is also referred to as Gross Domestic Product per barrel.

Geophysical Migration

Geophysical migration is the process by which geophysical events (changes in energy) are geometrically re-located in either space or time to the location the event occurred in the subsurface rather than the location that it was recorded at the surface, thereby creating an accurate image of the subsurface. This process is necessary to overcome the limitations of geophysical methods imposed by areas of complex geology, such as: faults, salt bodies, folding, etc. The end result is that the migrated image typically resolves these areas of complex geology much better than non-migrated images. Consequently, a form of migration is one of the standard data processing techniques for certain geophysical methods (seismic and ground penetrating radar). Computational migration needed for large datasets acquired today is extremely demanding on modern computers and is the subject of intense research, both within the geophysical industry as well as academic circles.

Need for Migration

Seismic waves are elastic waves that travel through the Earth with a finite velocity, which is governed by the material properties of the rocks. When the waves reach a rock unit with a different velocity some of the seismic energy is either absorbed, transmitted to the next layer, or reflected back towards the surface by the reflector (new rock layer). The reflected energy from these velocity contrasts arrive at the surface, and are recorded by Geophones that are placed at certain location away from the source of the waves. When a Geophysicist views the recorded energy from the Geophone, the geophysicist can only view the energy that was recorded at that location, but has no way of determining where it came from in the subsurface because (unlike Sonar), the seismic energy is not directed in the Earth. In the case that the source and receiver are at the same location (referred to as Zero-offset, where offset is the distance between the source and receiver), the geophysicist can guess at the event's location by (in the constant velocity world) using the relationship: $d = (1/2)r * t$ where d is the distance, r is the velocity (or rate of travel) and t is the measured time from the source to the receiver. In this case, the distance is halved because it can be assumed that it only took one-half the time from the reflector to the receiver, rather than the full measured time from the source to the receiver.

The result is troubling though, because it gives us only a singular value (a scalar). Therefore, the event could have come anywhere that has the same distance to the receiver. The result is that a half-sphere (a half-sphere, because we can ignore all possibilities that occur above the surface as unreasonable) represents all the possible locations that the event could have come from. This leads to zero-offset migration, wherein it can be assumed that the event came from somewhere on the sphere at that distance.

Zero-offset data is important to a geophysicist because the migration operation is much simpler, and can be represented by spherical surfaces. When data is acquired at offsets, the sphere becomes an ellipsoid, which is much more complex to represent (both geometrically, as well as computationally). Therefore, geophysicists would like to get their data to zero-offset before migrating it, because zero-offset migration is faster.

Zero-offset migration is more commonly referred to as Post-stack migration, wherein zero-offset traces are moved about in an image. Typically post-stack migration does a good job at imaging, but may fail in areas of complex geology due to some broad simplifications in the algorithms. Also, it is important to note, that in practical surveys data is almost always collected at offset, and then converted to zero-offset data. In doing so, much information is discarded for simplicity's sake.

For all intents and purposes, complex geology is defined as anywhere there is an abrupt or sharp contrast in lateral and/or vertical velocity (e.g. a sudden change in rock type or lithology which causes a sharp change in seismic wave velocity). Some examples of what a geophysicist considers complex geology are: faulting, folding, (some) fracturing, salt bodies, and unconformities. In these situations a form of migration is used that is referred to as pre-stack migration (PreSM), in which all traces are migrated before being moved to zero-offset. Consequently, much more information is used, which results in a much better image, along with the fact that PreSM honors velocity changes more accurately than post-stack migration.

Graphical Migration

The simplest form of migration is that of graphical migration. Graphical migration assumes a constant velocity world and zero-offset data, in which a geophysicist draws spheres or circles from the receiver to the event location for all events. The intersection of the circles then form the reflector's "true" location in time or space. An example of such can be seen in.

Types of Migration

Migration operators can be applied in both the time and space domains. The main difference between the two is how well the operator respects the velocity model. Time migrations typically use average values for velocity in computations, and as a result tend to poorly image overly complex areas. Time migrations are quick, and sufficient for areas with minor structural complexity. Depth migrations use the full velocity model, which leads to a (typically) much better image than time migration. However, depth migrations are substantially more expensive computationally than time migrations.

Time Migration

There are many formulations of Time migrations. Some of the more famous time migrations are: Stolt migration, Gazdag, and Finite-difference migration.

Depth Migration

There are many formulations of Depth migrations. Some of the more prominent depth migrations are: Kirchoff migration, Reverse Time Migration (RTM), Gaussian Beam Migration, and Wave-equation migration.

Technical Details

Migration of seismic data is the correction of the flat-geological-layer assumption by a numerical, grid-based spatial convolution of the seismic data to account for dipping events (where geological layers are not flat). There are many approaches, such as the popular Kirchoff migration, but it is generally accepted that processing large spatial sections (apertures) of the data at a time introduces fewer errors, and that depth migration is far superior to time migration with large dips and with complex salt bodies.

Basically, it repositions/moves the energy (seismic data) from the recorded locations to the locations with the correct common midpoint (CMP). While the seismic data is received at the proper locations originally (according to the laws of nature), these locations do not correspond with the assumed CMP for that location. Though stacking the data without the migration corrections yields a somewhat inaccurate picture of the subsurface, migration is preferred for better most imaging recorder to drill and maintain oilfields. This process is a central step in the creation of an image of the subsurface from active source seismic data collected at the surface, seabed, boreholes, etc., and therefore is used on industrial scales by oil and gas companies and their service providers on digital computers.

Explained in another way, this process attempts to account for wave dispersion from dipping reflectors and also for the spatial and directional seismic wave speed (heterogeneity) variations, which cause wavefields (modelled by ray paths) to bend, wave fronts to cross (caustics), and waves to be recorded at positions different from those that would be expected under straight ray or other simplifying assumptions. Finally, this process often attempts to also preserve and extract the formation interface reflectivity information imbedded in the seismic data amplitudes, so that they can be used to reconstruct the elastic properties of the geological formations (amplitude preservation, seismic inversion). There are a variety of migration algorithms, which can be classified by their output domain into the broad categories of Time Migration or Depth Migration, and Pre-Stack Migration or Post-Stack migration (orthogonal) techniques. Depth migration begins with time data converted to depth data by a spatial geological velocity profile. Post-Stack migration begins with seismic data which has already been stacked, and thus already lost valuable velocity analysis information.

Chapter 7

Oil Platform

An oil platform, also referred to as an offshore platform or, somewhat incorrectly, oil rig, is a làrge structure with facilities to drill wells, to extract and process oil and natural gas, and to temporarily store product until it can be brought to shore for refining and marketing. In many cases, the platform contains facilities to house the workforce as well.

Depending on the circumstances, the platform may be fixed to the ocean floor, may consist of an artificial island, or may float.

Remote subsea wells may also be connected to a platform by flow lines and by umbilical connections; these subsea solutions may consist of single wells or of a manifold centre for multiple wells.

History

Around 1891 the first submerged oil wells were drilled from platforms built on piles in the fresh waters of the Grand Lake St. Marys (a.k.a. Mercer County Reservoir) in Ohio. The wide but shallow reservoir was built from 1837 to 1845 to provide water to the Miami and Erie Canal.

Around 1896 the first submerged oil wells in salt water were drilled in the portion of the Summerland field extending under the Santa Barbara Channel in California. The wells were drilled from piers extending from land out into the channel.

Other notable early submerged drilling activities occurred on the Canadian side of Lake Erie in the 1900s and Caddo Lake in Louisiana in the 1910s. Shortly thereafter, wells were drilled in tidal zones along the Gulf Coast of Texas and Louisiana. The Goose Creek field near

Baytown, Texas is one such example. In the 1920s drilling was done from concrete platforms in Lake Maracaibo, Venezuela. The oldest subsea well recorded in Infield's offshore database is the Bibi Eibat well which came on stream in 1923 in Azerbaijan. Landfill was used to raise shallow portions of the Caspian Sea.

In the early 1930s the Texas Company developed the first mobile steel barges for drilling in the brackish coastal areas of the gulf.

In 1937 Pure Oil Company (now part of Chevron Corporation) and its partner Superior Oil Company (now part of ExxonMobil Corporation) used a fixed platform to develop a field in 14 feet (4.3 m) of water, one mile (1.6 km) offshore of Calcasieu Parish, Louisiana.

In 1946, Magnolia Petroleum Company (now part of ExxonMobil) erected a drilling platform in 18 ft (5.5 m) of water, 18 miles off the coast of St. Mary Parish, Louisiana.

In early 1947 Superior Oil erected a drilling/production platform in 20 ft (6.1 m) of water some 18 miles off Vermilion Parish, Louisiana. But it was Kerr-McGee Oil Industries (now Anadarko Petroleum Corporation), as operator for partners Phillips Petroleum (ConocoPhillips) and Stanolind Oil & Gas (BP), that completed its historic Ship Shoal Block 32 well in October 1947, months before Superior actually drilled a discovery from their Vermilion platform farther offshore. In any case, that made Kerr-McGee's well the first oil discovery drilled out of sight of land.

The Thames Sea Forts of World War II are considered the direct predecessors of modern offshore platforms. Having been pre-constructed in a very short time, they were then floated to their location and placed on the shallow bottom of the Thames estuary.

Types

Larger lake- and sea-based offshore platforms and drilling rigs are some of the largest moveable man-made structures in the world. There are several types of oil platforms and rigs:

Fixed Platform

A fixed platform is a type of offshore platform used for the production of oil or gas. These platforms are built on concrete and/or steel legs anchored directly onto the seabed, supporting a deck with space for drilling rigs, production facilities and crew quarters. Such platforms are, by virtue of their immobility, designed for very long term use (for instance the Hibernia platform). Various types of structure are used, steel jacket, concrete caisson, floating steel and even floating concrete. Steel jackets are vertical sections made of tubular steel members, and are usually piled into the seabed. Concrete caisson structures, pioneered by the Condeep concept, often have in-built oil storage in tanks below the sea surface and these tanks were often used as a flotation capability, allowing them to be built close to shore (Norwegian fjords and Scottish firths are popular because they are sheltered and deep enough) and then floated to their final position where they are sunk to the seabed. Fixed platforms are economically feasible for installation in water depths up to about 1,700 feet (520 m).

Compliant Tower

A compliant tower (CT) is a fixed rig structure normally used for the offshore production of oil or gas. The rig consists of narrow, flexible (*compliant*) towers and a piled foundation supporting a conventional deck for drilling and production operations. Compliant towers are designed to sustain significant lateral deflections and forces, and are typically used in water depths ranging from 1,500 and 3,000 feet (450 and 900 m). At present the deepest is Baldpate in 580 m of water.

With the use of flex elements such as flex legs or axial tubes, resonance is reduced and wave forces are de-amplified. This type of rig structure can be configured to adapt to existing fabrication and installation equipment.

Compared with floating systems, such as Tension-leg platforms and SPARs, the production risers are conventional and are subjected to less structural demands and flexing. This flexibility allows it to operate in much deeper water, as it can 'absorb' much of the pressure exerted on it by the wind and sea. Despite its flexibility, the compliant tower system is strong enough to withstand hurricane conditions.

The first tower emerged in the early 1980s with the installation of Exxon's Lena oil platform.

Launching of Benguela Belize Base Tower

On April 15, 2005 the Base Tower of Benguela Belize was launched from the barge H-851. The Base Tower makes up the first 250 metres of the more than 400 metre high tower. The crane vessel that is partly visible is the *Thialf*. This is the heaviest compliant tower installed to date and the first outside of the Gulf of Mexico.

Semi-submersible Platform

These platforms have hulls (columns and pontoons) of sufficient buoyancy to cause the structure to float, but of weight sufficient to keep the structure upright. Semi-submersible platforms can be moved from place to place; can be ballasted up or down by altering the amount of flooding in buoyancy tanks; they are generally anchored by combinations of chain, wire rope or polyester rope, or both, during drilling or production operations, or both, though they can also be kept in place by the use of dynamic positioning. Semi-submersibles can be used in water depths from 200 to 10,000 feet (60 to 3,000 m).

Jackup Rig

A jack-up rig is a type of mobile platform that is able to stand still on the sea floor, resting on a number of supporting legs. The most popular designs use 3 independent legs, although some jackups have 4 legs or more. On "mat-type" jackups the legs are connected to a submerged hull.

Operation

A jackup is a floating barge fitted with long support legs that can be raised or lowered. The jackup is maneuvered (self-propelled or by towing) into location with its legs up and the hull floating on the water. Upon arrival at the work location, the legs are jacked down onto the seafloor.

Then "preloading" takes place, where the weight of the barge and additional ballast water are used to drive the legs securely into the seabottom so they will not penetrate further while operations are carried out. After preloading, the jacking system is used to raise the entire barge above the water to a predetermined height or "air gap", so that wave, tidal and current loading acts only on the relatively slender legs and not on the barge hull.

Modern jacking systems use a rack and pinion gear arrangement where the pinion gears are driven by hydraulic or electric motors and the rack is affixed to the legs.

Jackup rigs can only be placed in relatively shallow waters, generally less than 400 feet (120 m) of water. However, a specialized class of jackup rigs known as premium or ultra-premium jackups are known to have operational capability in water depths ranging from 500 to 625 feet.

Types

Drilling Platforms

This type of rig is commonly used in connection with oil and/or natural gas drilling. There are more jackup rigs in the worldwide offshore rig fleet than any other type of mobile offshore drilling rig. Other types of offshore rigs include semi-submersibles (which float on pontoon-like structures) and drillships, which are ship-shape vessels with rigs mounted in the centers.

These rigs drill through holes in the drillship hulls, known as moon pools.

Barges

Jackup rigs also refer to specialized barges that are similar to an oil and gas platform but are used as a base for servicing other structures such as offshore wind turbines, long bridges, and drilling platforms.

Tension-leg Platform

A Tension-leg platform or Extended Tension Leg Platform (ETLP) is a vertically moored floating structure normally used for the offshore production of oil or gas, and is particularly suited for water depths greater than 300 metres (about 1000 ft) and less than 1500 meters (about 4900 ft). Use of tension-leg platforms has also been proposed for wind turbines.

The platform is permanently moored by means of tethers or tendons grouped at each of the structure's corners. A group of tethers is called a tension leg. A feature of the design of the tethers is that they have relatively high axial stiffness (low elasticity), such that virtually all vertical motion of the platform is eliminated. This allows the platform to have the production wellheads on deck (connected directly to the subsea wells by rigid risers), instead of on the seafloor. This allows a simpler well completion and gives better control over the production from the oil or gas reservoir, and easier access for downhole intervention operations.

TLP's have been in use since the early 1980s. The first Tension Leg Platform was built for Conoco's Hutton field in the North Sea in the early 1980s. The hull was built in the dry-dock at Highland Fabricator's Nigg yard in the north of Scotland, with the deck section built nearby at McDermott's yard at Ardersier. The two parts were mated in the Moray Firth in 1984.

Larger TLP's will normally have a full drilling rig on the platform with which to drill and intervene on the wells. The smaller TLPs may have a workover rig, or in a few cases no production wellheads located on the platform at all.

The deepest (E)TLPs measured from the sea floor to the surface are:

- 4,674 ft (1,425 m) Magnolia ETLP. Its total height is some 5,000 feet (1,500 m).
- 4,300 ft (1,300 m) Marco Polo TLP
- 4,250 ft (1,300 m) Neptune TLP
- 3,863 ft (1,177 m) Kizomba A TLP
- 3,800 ft (1,200 m) Ursa TLP. Its height above surface is 485 ft (148 m) making a total height of 4,285 ft (1,306 m).
- 3,350 ft (1,020 m) Allegheny TLP
- 3,300 ft (1,000 m) W. Seno A TLP.

Use for Wind Turbines

The Massachusetts Institute of Technology and the National Renewable Energy Laboratory first explored the concept of TLPs for offshore wind turbines in September 2006. Earlier offshore wind turbines cost more to produce, stood on towers dug deep into the ocean floor, were only possible in depths of at most 50 feet (15 m), and generated 1.5 megawatts for onshore units and 3.5 megawatts for conventional offshore setups. In contrast, TLP installation was calculated to cost a third as much. TLPs float, and researchers estimate they can operate in depths between 100 and 650 feet (200 m) and farther away from land, and they can generate 5.0 megawatts.

TLPs cost less to make and install because they are assembled onshore, are towed to their destination, and can be moved. Paul Sclavounos, an MIT professor of mechanical engineering and naval architecture who was involved in the design, said, "You don't pay anything to be buoyant."

Computer simulations project that in a hurricane TLPs would shift three to six feet and the turbine blades would cycle above wave peaks. MIT and NREL researchers say dampers could be used to reduce motion in the event of a natural disaster.

MIT and NREL researchers plan to install a half-scale prototype south of Cape Cod. Sclavounos said, "We'd have a little unit sitting out there to show that this thing can float and behave the way we're saying it will."

Gravity-based Structure

A gravity-based structure (GBS) is a support structure held in place by gravity. A common application for a GBS is an offshore oil platform. These structures are often constructed in fjords since their protected area and sufficient depth are very desirable for construction.

A GBS intended for use as an offshore oil platform is constructed of steel reinforced concrete, often with tanks or cells which can be used to control the buoyancy of the finished GBS. When completed, a GBS is towed to its intended location and sunk. The platform structure which a GBS supports is called the topsides.

Gravity-based structures are also widely used for offshore wind power plants. By the end of 2010, 14 of the world's offshore wind farms were supported by Gravity-based structures. The GBS are suited for water depths up to 20 meters.

The deepest registered offshore wind farm with Gravity-based structures is Thornton Bank 1, Belgium, with a depth up to 27.5 meters. However, as offshore wind power plants are growing in size and moving towards deeper waters, the GBS is not considered competitive in comparison with other support structures.

Spar (Platform)

A spar, named for logs used as buoys in shipping and moored in place vertically, is a type of floating oil platform typically used in very deep waters. Spar production platforms have been developed as an alternative to conventional platforms.

A spar platform consists of a large-diameter, single vertical cylinder supporting a deck.

The first spar platform in the Gulf of Mexico was installed in September 1996.

The world's deepest platform is the Perdido in the Gulf of Mexico, floating in 2,438 meters of water. It is operated by Royal Dutch Shell and was built at a cost of $3 billion.

Normally Unmanned Installation

A Normally Unmanned Installation (NUI) is a type of automated offshore Oil/Gas platform designed to be primarily operated remotely, without the constant presence of personnel. These generally were characterized by their small size, often consisting of just a well bay with a helipad on top. They are often a compromise of providing the convenience of surface wellheads, which are easier to build and maintain, while avoiding the high operating costs of a full production platform.

They are generally only used in shallower water, where constructing many small NUIs is a relatively easy and cheap option as compared to the cost of using subsea wells.

This can be seen in the Southern North Sea where large numbers of wells are on smaller NUIs, compared with the more northern areas of the continental shelf where fewer larger platforms and subsea sites are the norm.

NUIs are commonly serviced from a nearby larger platform, e.g., Mungo serviced from Marnock. These installations will include an emergency shelter with essential food and water in order to provide a safe refuge in the event that weather or other considerations prevent a visiting crew from returning to base.

Regular visits may be made for routine maintenance and for smaller well work such as wireline operations. Anything larger requires a drilling rig to be brought in, but this is still an advantage over subsea wells, which require a drilling rig or light intervention vessel for any well intervention.In recent times NUI has become a phase within the decommissioning of previously manned offshore installations.

Platforms would generally be positively isolated from hydrocarbons and flushed clean to suit environmental issues. Platforms would then only be visited infrequently for integrity checks and maintenance of temporary equipment left for power requirements and safety functionality. The platform would then remain in this phase until further decommissioning activities are carried out to remove the topsides in modules or piece small, then remove the jacket.

Conductor Support System

Conductor support systems, also known as conductor supported systems or satellite platforms, are installations which are small unmanned platforms consisting of little more than a well bay, and a small process plant. They are designed to operate in conjunction with a static production platform which is connected to the platform by flow lines and/or by Umbilical cable.

Traditionally, these jacket-type structures have been installed and used in shallow to medium water depths of up to 40 – 60 meters. The conductor supported system use its inherited strength of the well conductors to support both the wells and the topside structure.

The conductor supported system is particularly suited to areas with more benign environmental conditions. The well conductors act as both piles and conductors.

These are drilled and installed using a drilling jackup rig using conventional drilling/lifting techniques.

Particularly Large Examples

The Petronius Platform is a compliant tower in the Gulf of Mexico, which stands 2,000 feet (610 m) above the ocean floor. It is one of the world's tallest structures.

The Hibernia platform in Canada is the world's largest (in terms of weight) offshore platform, located on the Jeanne D'Arc basin, in the Atlantic Ocean off the coast of Newfoundland. This *gravity base structure* (GBS), which sits on the ocean floor, is 364 feet (111 m) high and has storage capacity for 1.3 million barrels (210,000 m^3) of crude oil in its 278.8-foot (85.0 m) high caisson.

The platform acts as a small concrete island with serrated outer edges designed to withstand the impact of an iceberg.

The GBS contains production storage tanks and the remainder of the void space is filled with ballast with the entire structure weighing in at 1.2 million tons.

Royal Dutch Shell is currently developing the first Floating Liquefied Natural Gas (FLNG) facility, which will be situated approximately 200km off the coast of Western Australia and is due for completion around 2017. When finished, it will be the largest floating offshore facility. It is expected to be approximately 488m long and 74m wide with displacement of around 600,000t when fully ballasted.

Maintenance and Supply

A typical oil production platform is self-sufficient in energy and water needs, housing electrical generation, water desalinators and all of the equipment necessary to process oil and gas such that it can be either delivered directly onshore by pipeline or to a floating platform or tanker loading facility, or both.

Elements in the oil/gas production process include wellhead, production manifold, production separator, glycol process to dry gas, gas compressors, water injection pumps, oil/gas export metering and main oil line pumps.

Larger platforms assisted by smaller ESVs (emergency support vessels) like the British Iolair that are summoned when something has gone wrong, *e.g.* when a search and rescue operation is required.

During normal operations, PSVs (platform supply vessels) keep the platforms provisioned and supplied, and AHTS vessels can also supply them, as well as tow them to location and serve as standby rescue and firefighting vessels.

Crew

Essential Personnel

Not all of the following personnel are present on every platform. On smaller platforms, one worker can perform a number of different jobs.

The following also are not names officially recognized in the industry:

- OIM (offshore installation manager) who is the ultimate authority during his/her shift and makes the essential decisions regarding the operation of the platform;
- operations team leader (OTL);
- offshore operations engineer (OOE) who is the senior technical authority on the platform;
- PSTL or operations coordinator for managing crew changes;
- dynamic positioning operator, navigation, ship or vessel maneuvering (MODU), station keeping, fire and gas systems operations in the event of incident;
- second mate to meet manning requirements of flag state, operates fast rescue craft, cargo operations, fire team leader;
- third mate to meet manning requirements of flag state, operate fast rescue craft, cargo operations, fire team leader;
- ballast control operator to operate fire and gas systems;
- crane operators to operate the cranes for lifting cargo around the platform and between boats;
- scaffolders to rig up scaffolding for when it is required for workers to work at height;
- coxswains to maintain the lifeboats and manning them if necessary;
- control room operators, especially FPSO or production platforms;
- catering crew, including people tasked with performing essential functions such as cooking, laundry and cleaning the accommodation;
- production techs to run the production plant;
- helicopter pilot(s) living on some platforms that have a helicopter based offshore and transporting workers to other platforms or to shore on crew changes;
- maintenance technicians (instrument, electrical or mechanical).

Incidental Personnel

Drill crew will be on board if the installation is performing drilling operations. A drill crew will normally comprise:

- Toolpusher
- Driller
- Roughnecks
- Roustabouts
- Company man
- Mud engineer
- Derrickhand
- Geologist.

Well services crew will be on board for well work. The crew will normally comprise:

- Well services supervisor
- Wireline or coiled tubing operators
- Pump operator.

Risks

The nature of their operation — extraction of volatile substances sometimes under extreme pressure in a hostile environment — means risk; accidents and tragedies occur regularly. The U.S. Minerals Management Service reported 69 offshore deaths, 1,349 injuries, and 858 fires and explosions on offshore rigs in the Gulf of Mexico from 2001 to 2010. In July 1988, 167 people died when Occidental Petroleum's Piper Alpha offshore production platform, on the Piper field in the UK sector of the North Sea, exploded after a gas leak.

The resulting investigation conducted by Lord Cullen and publicized in the first Cullen Report was highly critical of a number of areas, including, but not limited to, management within the company, the design of the structure, and the Permit to Work System. The report was commissioned in 1988, and was delivered November 1990. The accident greatly accelerated the practice of providing living accommodations on separate platforms, away from those used for extraction. However, this was in itself a hazardous environment. In March 1980, the 'flotel' (floating hotel) platform *Alexander L. Kielland* capsized in a storm in the North Sea with the loss of 123 lives.

In 2001, *Petrobras 36* in Brazil exploded and sank five days later, killing 11 people.

Given the number of grievances and conspiracy theories that involve the oil business, and the importance of gas/oil platforms to the economy, platforms in the United States are believed to be potential terrorist targets. Agencies and military units responsible for maritime counterterrorism in the US (Coast Guard, Navy SEALs, Marine Recon) often train for platform raids.

On April 21, 2010, the *Deepwater Horizon* platform, 52 miles offshore of Venice, Louisiana, (property of Transocean and leased to BP) exploded, killing 11 people, and sank two days later. The resulting undersea gusher, conservatively estimated to exceed 20 million US gallons (76,000 m^3) as of early June, 2010, became the worst oil spill in US history, eclipsing the Exxon Valdez oil spill.

Ecological Effects

In British waters, the cost of removing all platform rig structures entirely was estimated in 1995 at £1.5 billion, and the cost of removing all structures including pipelines—called a "clean sea" approach—at £3 billion.

In the United States, Marine Biologist Milton Love has proposed that oil platforms off the California coast be retained as artificial reefs, instead of being dismantled (at great cost), because he has found them to be havens for many of the species of fish which are otherwise declining in the region, in the course of 11 years of research. Love is funded mainly by government agencies, but also in small part by the California Artificial Reef Enhancement Program.

NOAA has said it is considering this course of action, but wants money to study the effects of the rigs in detail. Divers have been used to assess the fish populations surrounding the platforms. In the Gulf of Mexico, more than 200 platforms have been similarly converted.

Deepest Oil Platforms

The world's deepest oil platform is the floating Perdido which is a spar platform in the Gulf of Mexico in a water depth of 2,438 metres (7,999 ft).

Non-floating compliant towers and fixed platforms, by water depth:

- Petronius Platform, 531 m (1,742 ft)
- Baldpate Platform, 502 m (1,647 ft)
- Bullwinkle Platform, 413 m (1,355 ft)
- Pompano Platform, 393 m (1,289 ft)
- Benguela-Belize Lobito-Tomboco Platform, 390 m (1,280 ft)

- Tombua Landana Platform, 366 m (1,201 ft)
- Harmony Platform, 366 m (1,201 ft)
- Troll A Platform, 303 m (994 ft)
- Gulfaks C Platform, 217 m (712 ft).

Devil's Tower (Oil Platform)

Devil's Tower is located in block 773 of the Mississippi Canyon region of the Gulf of Mexico. Until 2010 it was the record holder for being located in the deepest water for a production truss spar (this record was broken by Royal Dutch Shell's Perdido platform). It was completed at the end of 2003. Originally it was built for and operated by Dominion Exploration which sold its Gulf of Mexico holdings to Eni which now operates the platform. The platform is owned by Williams Energy (NYSE: WMB)

The facility was built by J. Ray McDermott and is located 5,610 feet (1,710 m) above the seabed.

Drill Floor

The Drill Floor is the heart of any drilling rig and is also known as the *pad*. This is the area where the drill string begins its trip into the earth. It is traditionally where joints of pipe are assembled, as well as the BHA (bottom hole assembly), drilling bit, and various other tools. This is the primary work location for roughnecks and the driller. The drill floor is located directly under the derrick.

Chapter 8

Platform Supply Vessel

A Platform supply vessel (often abbreviated as PSV) is a ship specially designed to supply offshore oil platforms. These ships range from 20 to 100 meters in length and accomplish a variety of tasks. The primary function for most of these vessels is transportation of goods and personnel to and from offshore oil platforms and other offshore structures.

In the recent years a new generation of Platform Supply Vessel entered the market, usually equipped with Class 1 or Class 2 Dynamic Positioning System.

Capabilities

Cargo

A primary function of a platform supply vessel is to transport supplies to the oil platform and return other cargoes to shore. Cargo tanks for *drilling mud*, pulverized cement, diesel fuel, potable and non-potable water, and chemicals used in the drilling process comprise the bulk of the cargo spaces. Fuel, water, and chemicals are almost always required by oil platforms. Certain other chemicals must be returned to shore for proper recycling or disposal, however, crude oil product from the rig is usually not a supply vessel cargo.

Support

Common and specialty tools are carried on the large decks of these vessels. Most carry a combination of deck cargoes and bulk cargo in tanks below deck. Many ships are constructed (or re-fitted) to accomplish a particular job. Some of these vessels are equipped with a firefighting capability and fire monitors for fighting platform fires. Some vessels are equipped with oil containment and recovery equipment to assist in the cleanup of a spill at sea. Other vessels are equipped with tools, chemicals and personnel to "work-over" existing oil wells for the purpose of increasing the wells' production.

Vessel Crews

Crews on the these ships can number up to 20 crewmembers.

Daily Operations

Crews normally sign on to work and live aboard the ship an extended period of time (3 – 6 weeks and often more). This is followed by an extended period of time off, often 2–4 weeks depending on the ship owner/operator the number of weeks or months aboard. Work details on platform supply vessels, like many ships, are organized into shifts of up to 12 hours.

Living aboard the ship, each crew member and worker will have a shift, lasting some portion of a 24 hour day. Most supply vessels are provided with a "bridge" area for navigating and operating the ship, machinery spaces, living quarters, and an area for cooking and eating. Some have built in work areas, and common areas for entertainment. The large main deck area is sometimes utilized for portable housing.

Living quarters consist of a bunk area, lockers, and spaces for storing personal items. Living areas are provided with wash basins, showers and toilets. Officers living quarters are sometimes outfitted with a small work desk, private sinks, showers and toilets.

The "galley" or cooking and eating areas aboard ship will be stocked with enough grocery items to last for the intended voyage. A walk-in size cooler and freezer, a commercial stove and oven, deep sinks, storage and counterspace will be available for the persons doing the cooking. The eating area will have coffee makers, toasters, microwave ovens, cafeteria style seating, and other amenities needed to feed a hard working crew.

Extraction of Petroleum

The extraction of petroleum is the process by which usable petroleum is extracted and removed from the earth.

Locating the Oil Field

Geologists use seismic surveys to search for geological structures that may form oil reservoirs. The "classic" method includes making an underground explosion nearby and observing the seismic response that provides information about the geological structures under the ground. However, "passive" methods that extract information from naturally-occurring seismic waves are also known.

Other instruments such as gravimeters and magnetometers are also sometimes used in the search for petroleum. Extracting crude oil normally starts with drilling wells into the underground reservoir. When an oil well has been tapped, a geologist (known on the rig as the "mudlogger") will note its presence. Such a "mudlogger" is known to be sitting on the rig.

Historically, in the USA, some oil fields existed where the oil rose naturally to the surface, but most of these fields have long since been used up, except in certain places in Alaska. Often many wells (called *multilateral wells*) are drilled into the same reservoir, to ensure that the extraction rate will be economically viable. Also, some wells (*secondary wells*) may be used to pump water, steam, acids or various gas mixtures into the reservoir to raise or maintain the reservoir pressure, and so maintain an economic extraction rate.

Drilling

The oil well is created by drilling a hole into the earth with an oil rig. A steel pipe (casing) is placed in the hole, to provide structural integrity to the newly drilled wellbore. Holes are then made in the base of the well to enable oil to pass into the bore. Finally a collection of valves called a "Christmas Tree" is fitted to the top, the valves regulating pressures and controlling flows.

Oil Extraction and Recovery

Primary Recovery

During the *primary recovery stage*, reservoir drive comes from a number of natural mechanisms. These include: natural water displacing oil downward into the well, expansion of the natural gas at the top of the reservoir, expansion of gas initially dissolved in the crude oil, and gravity drainage resulting from the movement of oil within the reservoir from the upper to the lower parts where the wells are located. Recovery factor during the primary recovery stage is typically 5-15%.

While the underground pressure in the oil reservoir is sufficient to force the oil to the surface, all that is necessary is to place a complex arrangement of valves (the Christmas tree) on the well head to connect the well to a pipeline network for storage and processing.

Secondary Recovery

Over the lifetime of the well the pressure will fall, and at some point there will be insufficient underground pressure to force the oil to the surface. After natural reservoir drive diminishes, *secondary recovery* methods are applied.

They rely on the supply of external energy into the reservoir in the form of injecting fluids to increase reservoir pressure, hence replacing or increasing the natural reservoir drive with an artificial drive. Sometimes pumps, such as beam pumps and electrical submersible pumps (ESPs), are used to bring the oil to the surface.

Other secondary recovery techniques increase the reservoir's pressure by water injection, natural gas reinjection and gas lift, which injects air, carbon dioxide or some other gas into the bottom of an active well, reducing the overall density of fluid in the wellbore.

Typical recovery factor from water-flood operations is about 30%, depending on the properties of oil and the characteristics of the reservoir rock. On average, the recovery factor after primary and secondary oil recovery operations is between 35 and 45%.

Tertiary Recovery

Tertiary, or enhanced oil recovery methods increase the mobility of the oil in order to increase extraction.

Thermally enhanced oil recovery methods (TEOR) are tertiary recovery techniques that heat the oil, thus reducing its viscosity and making it easier to extract.

Steam injection is the most common form of TEOR, and is often done with a cogeneration plant. In this type of cogeneration plant, a gas turbine is used to generate electricity and the waste heat is used to produce steam, which is then injected into the reservoir. This form of recovery is used extensively to increase oil extraction in the San Joaquin Valley, which has very heavy oil, yet accounts for 10% of the United States' oil extraction. In-situ burning is another form of TEOR, but instead of steam, some of the oil is burned to heat the surrounding oil.

Occasionally, surfactants (detergents) are injected to alter the surface tension between the water and oil in the reservoir, mobilizing oil which would otherwise remain in the reservoir as residual oil.

Another method to reduce viscosity is carbon dioxide flooding. Tertiary recovery allows another 5% to 15% of the reservoir's oil to be recovered.

Tertiary recovery begins when secondary oil recovery isn't enough to continue adequate extraction, but only when the oil can still be extracted profitably.

This depends on the cost of the extraction method and the current price of crude oil. When prices are high, previously unprofitable wells are brought back into use and when they are low, extraction is curtailed.

Microbial treatments is another tertiary recovery method. Special blends of the microbes are used to treat and break down the hydrocarbon chain in oil thus making the oil easy to recover as well as being more economic versus other conventional methods.

In some states, such as Texas, there are tax incentives for using these microbes in what is called a secondary tertiary recovery. Very few companies supply these, however companies like Bio Tech, Inc. have proven very successful in waterfloods across Texas.

Recovery Rates

The amount of oil that is recoverable is determined by a number of factors including the permeability of the rocks, the strength of natural drives (the gas present, pressure from adjacent water or gravity), and the viscosity of the oil.

When the reservoir rocks are "tight" such as shale, oil generally cannot flow through but when they are permeable such as in sandstone, oil flows freely.

The flow of oil is often helped by natural pressures surrounding the reservoir rocks including natural gas that may be dissolved in the oil, natural gas present above the oil, water below the oil and the strength of gravity.

Oils tend to span a large range of viscosity from liquids as light as gasoline to heavy as tar. The lightest forms tend to result in higher extraction rates.

Petroleum engineering is the discipline responsible for evaluating which well locations and recovery mechanisms are appropriate for a reservoir and for estimating recovery rates and oil reserves prior to actual extraction.

Carbon Dioxide Flooding

Carbon dioxide (CO_2) flooding is a process whereby carbon dioxide is injected into an oil reservoir in order to increase output when extracting oil.

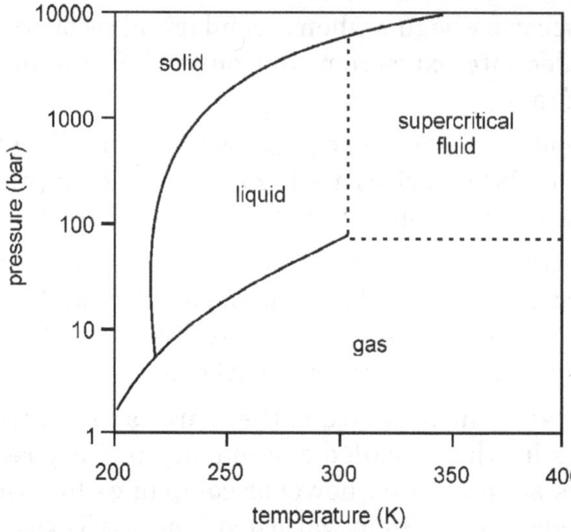

***Figure:** Carbon dioxide pressure-temperature phase diagram*

When a reservoir's pressure is depleted through primary and secondary production, Carbon Dioxide flooding can be an ideal tertiary recovery method. It is particularly effective in reservoirs deeper than 2,000 ft., where CO_2 will be in a supercritical state, with API oil gravity greater than 22–25° and remaining oil saturations greater than 20%. It should also be noted that Carbon dioxide flooding is not affected by the lithology of the reservoir area but simply by the reservoir characteristics. Carbon dioxide flooding works on the premise that by injecting CO_2 into the reservoir, the viscosity of any hydrocarbon will be reduced and hence will be easier to sweep to the production well.

If a well has been produced before and has been designated suitable for CO_2 flooding, the first thing to do is to restore the pressure within the reservoir to one suitable for production. This is done by injecting water (with the production well shut off) which will restore pressure within the reservoir to a suitable pressure for CO_2 flooding. Once the reservoir is at this pressure, the next step is to inject the CO_2 into the same injection wells used to restore pressure.

The CO_2 gas is forced into the reservoir and is required to come into contact with the oil. This creates this miscible zone that can be moved easier to the production well. Normally the CO_2 injection is alternated with more water injection and the water acts to sweep the oil towards the production zone.

The Weyburn oil field is a famous example where this method is applied in financially interesting conditions. CO_2 flooding is the second most common tertiary recovery technique and is used in facilities around the world. In the scope of global warming, it is an available method to curb CO_2 emissions.

Casing (Borehole)

Casing is large diameter pipe that is assembled and inserted into a recently drilled section of a borehole and typically held into place with cement.

Purpose

Casing that is cemented in place aids the drilling process in several ways:

- Prevent contamination of fresh water well zones.
- Prevent unstable upper formations from caving-in and sticking the drill string or forming large caverns.
- Provides a strong upper foundation to use high-density drilling fluid to continue drilling deeper.
- Isolates different zones, that may have different pressures or fluids - known as zonal isolation, in the drilled formations from one another.
- Seals off high pressure zones from the surface, avoiding potential for a blowout
- Prevents fluid loss into or contamination of production zones.
- Provides a smooth internal bore for installing production equipment.

A slightly different metal string, called production tubing, is often used without cement in the smallest casing of a well completion to contain production fluids and convey them to the surface from an underground reservoir.

Design

In the planning stages of a well a drilling engineer, usually with input from geologists and others, will pick strategic depths at which the hole will need to be cased in order for drilling to reach the desired total depth. This decision is often based on subsurface data such as formation pressures, strengths, and makeup, and is balanced against the cost objectives and desired drilling strategy.

With the casing set depths determined, hole sizes and casing sizes must follow. The hole drilled for each casing string must be large enough to easily fit the casing inside it, allowing room for cement between the outside of the casing and the hole. Also, the inside diameter of the first casing string must be large enough to fit the second bit that will continue drilling. Thus, each casing string will have a subsequently smaller diameter. The inside diameter of the final casing string (or penultimate one in some instances of a liner completion) must accommodate the production tubing and associated hardware such as packers, gas lift mandrels and subsurface safety valves.

Casing design for each size is done by calculating the worst condition that may be faced during drilling and production. Mechanical properties of designed pipes such as collapse resistance, burst pressure, and tensile strength must be sufficient for the worst conditions. Casing strings are supported by casing hangers that are set in the wellhead, which later will be topped with the Christmas tree. The wellhead usually is installed on top of the first casing string after it has been cemented in place.

Intervals

Typically, a well contains multiple intervals of casing successively placed within the previous casing run. The following casing interval is typically used in an oil or gas well:

- Conductor casing
- Surface casing
- Intermediate casing (optional)
- Production casing
- Production liner.

The conductor casing serves as a support during drilling operations, to flowback returns during drilling and cementing of the surface casing, and to prevent collapse of the loose soil near the surface. It can normally vary from sizes such as 18" to 30".

The purpose of surface casing is to isolate freshwater zones so that they are not contaminated during drilling and completion. Surface casing is the most strictly regulated due to these environmental concerns, which can include regulation of casing depth and cement quality. A typical size of surface casing is 13\! inches. Intermediate casing may be necessary on longer drilling intervals where necessary drilling mud weight to prevent blowouts may cause a hydrostatic pressure that can fracture deeper formations. Casing placement is selected so that the hydrostatic pressure of the drilling fluid remains between formation pore and fracture pressures.

The final interval is production casing. As with the casing intervals described above, the production casing string extends to the surface where it is hung off. As the smallest casing, it will form the outer boundary of the 'A' annulus, which may involve it being used for gas lift and well kills. A typical size is 9]! inches.

In order to reduce cost, a liner may be used which extends just above the shoe (bottom) of the previous casing interval and hung off downhole rather than at the surface. It may typically be 7", although many liners match the diameter of the production tubing. Few wells actually produce through casing, since producing fluids can corrode steel or form deposits such as asphaltenes or paraffins and the larger diameter can make flow unstable. Production tubing is therefore installed inside the last casing string and the tubing annulus is usually sealed at the bottom of the tubing by a packer. Tubing is easier to remove for maintenance, replacement, or for various types of workover operations. It is significantly lighter than casing and does not require a drilling rig to run in and out of hole; smaller "service rigs" are used for this purpose.

Cementing

Cementing is performed by circulating a cement slurry through the inside of the casing and out into the annulus through the casing shoe at the bottom of the casing string. In order to precisely place the cement slurry at a required interval on the outside of the casing, a plug is pumped with a displacement fluid behind the cement slurry column, which "bumps" in the casing shoe and prevents further flow of fluid through the shoe. This bump can be seen at surface as a pressure spike at the cement pump. To prevent the cement from flowing back into the inside of the casing, a float collar above the casing shoe acts as a check valve and prevents fluid from flowing up through the shoe from the annulus.

Coiled Tubing Umbilical

Coiled tubing umbilicals are a type of piping used in and gas wells, which can be used during interventions. They encapsulate multiple coiled tubing and electrical strings into a single string which can be deployed by one conventional unit. These technologies allow for the deployment of complex pumps on coiled tubing. Multiple strings in a casing may maintain a manageable shape for the tubing and eliminating the danger of ovalization.

Cyberbase

The Cyberbase is an ergonomic operator station that is used to control & monitor the drilling equipment that is used while drilling for oil/gas. It has two monitors that shows vital information and two control panels with joysticks and keypads. Made by National Oilwell Varco

Drill String

A drill string on a drilling rig is a column, or string, of drill pipe that transmits drilling fluid (via the mud pumps) and torque (via the kelly drive or top drive) to the drill bit.

The term is loosely applied as the assembled collection of the drill pipe, drill collars, tools and drill bit. The drill string is hollow so that drilling fluid can be pumped down through it and circulated back up the annulus (the void between the drill string and the casing/open hole).

Drill String Components

The drill string is typically made up of three sections:

- Bottom hole assembly (BHA)
- Transition pipe, which is often heavyweight drill pipe (HWDP)
- Drill pipe.

Bottom Hole Assembly (BHA)

The BHA is made up of a drill bit, which is used to break up the rock formations, drill collars, which are heavy, thick-walled tubes used to apply weight to the drill bit, and drilling stabilizers, which keep the assembly centered in the hole.

The BHA may also contain other components such as a downhole motor and Rotary steerable system, measurement while drilling (MWD), and logging while drilling (LWD) tools. The components are joined together using rugged threaded connections. Short "subs" are used to connect items with dissimilar threads.

Transition Pipe

Heavyweight drill pipe (HWDP) may be used to make the transition between the drill collars and drill pipe. The function of the HWDP is to provide a flexible transition between the drill collars and the drill pipe. This helps to reduce the number of fatigue failures seen directly above the BHA. A secondary use of HWDP is to add additional weight to the drill bit.

Drill Pipe

Drill pipe makes up the majority of the drill string back up to the surface. Each segment has a larger-diameter portion on each end containing a male ("pin") or female ("box") thread.

Running a Drill String

Most components in a drill string are manufactured in 31 foot lengths (range 2) although they can also be manufactured in 46 foot lengths (range 3). Each 31 foot component is referred to as a joint. Typically 2, 3 or 4 joints are joined together to make a stand. Modern onshore rigs are capable of handling ~90 ft stands (often referred to as a triple).

Pulling the drill string out of or running the drill string into the hole is referred to as tripping. Drill pipe, HWDP and collars are typically stood back in stands in the derrick if they are to be run back into the hole again after, say, changing the bit. The disconnect point ("break") is varied each subsequent round trip so that after three trips every connection has been broken apart and later made up again with fresh pipe dope applied.

Stuck Drill String

A stuck drill string can be caused by many situations.

- Packing-off due to cuttings settling back into the wellbore when circulation is stopped.
- Differentially when the formation pressure is too low, the wellbore pressure is too high or both, essentially pushing the pipe onto the wall of the wellbore.
- Keyhole sticking occurs mechanically as a result of pulling up into doglegs when tripping.
- Adhesion due to not moving it for a significant amount of time.

Once the tubular member is stuck, there are many techniques used to extract the pipe. The tools and expertise are normally supplied by an oilfield service company. Two popular tools and techniques are the oilfield jar and the surface resonant vibrator. Below is a history of these tools along with how they operate.

History of Jars

The mechanical success of cable tool drilling has greatly depended on a device called jars, invented by a spring pole driller, William Morris, in the salt well days of the 1830s. Little is known about Morris except for his invention and that he listed Kanawha County (now in West Virginia) as his address.

Morris received US 2243 for this unique tool in 1841 for artesian well drilling. Later, using jars, the cable tool system was able to efficiently meet the demands of drilling wells for oil.

The jars were improved over time, especially at the hands of the oil drillers, and reached the most useful and workable design by the 1870s, due to another US 78958 received in 1868 by Edward Guillod of Titusville, Pennsylvania, which addressed the use of steel on the jars' surfaces that were subject to the greatest wear. Many years later, in the 1930s, very strong steel alloy jars were made.

A set of jars consisted of two interlocking links which could telescope. In 1880 they had a play of about 13 inches such that the upper link could be lifted 13 inches before the lower link was engaged. This engagement occurred when the cross-heads came together.Today, there are two primary types, hydraulic and mechanical jars. While their respective designs are quite different, their operation is similar. Energy is stored in the drillstring and suddenly released by the jar when it fires.

Jars can be designed to strike up, down, or both. In the case of jarring up above a stuck bottomhole assembly, the driller slowly pulls up on the drillstring but the BHA does not move. Since the top of the drillstring is moving up, this means that the drillstring itself is stretching and storing energy. When the jars reach their firing point, they suddenly allow one section of the jar to move axially relative to a second, being pulled up rapidly in much the same way that one end of a stretched spring moves when released. After a few inches of movement, this moving section slams into a steel shoulder, imparting an impact load.

In addition to the mechanical and hydraulic versions, jars are classified as drilling jars or fishing jars. The operation of the two types is similar, and both deliver approximately the same impact blow, but the drilling jar is built such that it can better withstand the rotary and vibrational loading associated with drilling. Jars are designed to be reset by simple string manipulation and are capable of repeated operation or firing before being recovered from the well. Jarring effectiveness is determined by how rapidly you can impact weight into the jars.

When jarring without a compounder or accelerator you rely only on pipe stretch to lift the drill collars upwards after the jar releases to create the upwards impact in the jar. This accelerated upward movement will often be reduced by the friction of the working string along the sides of the well bore, reducing the speed of upwards movement of the drill collars which impact into the jar. At shallow depths jar impact is not achieved because of lack of pipe stretch in the working string.

When pipe stretch alone cannot provide enough energy to free a fish, compounders or accelerators are used. Compounders or accelerators are energized when you over pull on the working string and compress a compressible fluid through a few feet of stroke distance and at the same time activate the fishing jar. When the fishing jar releases the stored energy in the compounder/acclerator lifts the drill collars upwards at a high rate of speed creating a high impact in the jar.

System Dynamics of Jars

Jars rely on the principle of stretching a pipe to build elastic potential energy such that when the jar trips it relies on the masses of the drill pipe and collars to gain velocity and subsequently strike the anvil section of jar. This impact results in a force, or blow, which is converted into energy.

History of Surface Resonant Vibrators

The concept of using vibration to free stuck objects from a wellbore originated in the 1940s, and probably stemmed from the 1930s use of vibration to drive piling in the Soviet Union. The early use of vibration for driving and extracting piles was confined to low frequency operation; that is, frequencies less than the fundamental resonant frequency of the system and consequently, although effective, the process was only an improvement on conventional hammer equipment.

Early patents and teaching attempted to explain the process and mechanism involved, but lacked a certain degree of sophistication. In 1961, A. G. Bodine obtained US 2972380 that was to become the "mother patent" for oil field tubular extraction using sonic techniques. Mr. Bodine introduced the concept of resonant vibration that effectively eliminated the reactance portion of mechanical impedance, thus leading to the means of efficient sonic power transmission. Subsequently, Mr. Bodine obtained additional patents directed to more focused applications of the technology.

The first published work on this technique was outlined in a 1987 Society of Petroleum Engineers (SPE) paper presented at the International Association of Drilling Contractors in Dallas, Texas detailing the nature of the work and the operational results that were achieved. The cited work involving liner, tubing, and drill pipe extraction and was very successful. Reference Two presented at the Society of Petroleum Engineers Annual Technical Conference and Exhibition in Aneheim, Ca, November, 2007 explains the resonant vibration theory in more detail as well as its use in extracting long lengths of mud stuck tubulars.

System Dynamics of Surface Resonant Vibrators

Surface Resonant Vibrators rely on the principle of counter rotating eccentric weights to impart a sinusoidal harmonic motion from the surface into the work string at the surface. Reference Three (above) provides a full explanation of this technology. The frequency of rotation, and hence vibration of the pipe string, is tuned to the resonant frequency of the system.

The system is defined as the surface resonant vibrator, pipe string, fish and retaining media. The resultant forces imparted to the fish is based on the following logic:

- The delivery forces from the surface are a result of the static overpull force from the rig, plus the dynamic force component of the rotating eccentric weights
- Depending on the static overpull force component, the resultant force at the fish can be either tension or compression due to the sinusoidal force wave component from the oscillator
- Initially during startup of a vibrator, some force is necessary to lift and lower the entire load mass of the system. When the vibrator tunes to the resonant frequency of the system, the reactive load impedance cancels out to zero by virtue of the inductance reactance (mass of the system) equalling the compliance or stiffness reactance (elasticity of the tubular). The remaining impedance of the system, known as the resistive load impedance, is what is retaining the stuck pipe.
- During resonant vibration, a longitudinal sine wave travels down the pipe to the fish with an attendant pipe mass that is equal to a quarter wavelength of the resonant vibrating frequency.
- A phenomenon known as fluidization of soil grains takes place during resonant vibration whereby the granular material constraining the stuck pipe is transformed into a fluidic state that offers little resistance to movement of bodies through the media. In effect, it takes on the characteristics and properties of a liquid.
- During pipe vibration, Dilation and Contraction of the pipe body, known as Poisson's ratio, takes place such that that when the stuck pipe is subjected to axial strain due to stretching, its diameter will contract. Similarly, when the length of pipe is compressed, its diameter will expand. Since a length of pipe undergoing vibration experiences alternate tensile and

compressive forces as waves along its longitudinal axis (and therefore longitudinal strains), its diameter will expand and contract in unison with the applied tensile and compressive waves. This means that for alternate moments during a vibration cycle the pipe may actually be physically free of its bond.

Annulus (Oil Well)

The annulus (yellow area in diagram) of an oil well refers to any void between any piping, tubing or casing and the piping, tubing, or casing immediately surrounding it. It is named after the corresponding geometric concept. The presence of an annulus gives the ability to circulate fluid in the well, provided that excess drill cuttings have not accumulated in the annulus, preventing fluid movement and possibly sticking the pipe in the borehole.

For a new well in the process of being drilled, this would be the void between the drill string and the formation being drilled. An easy way to visualise this would be to stand a straw (purple in diagram) straight up in the centre of a glass of water. All of the water in between the straw and the sides of the glass would be the annulus (yellow area in diagram), with the straw itself representing the drill string and the sides of the glass representing the formation. While drilling, drilling fluid is pumped down the inside of the drill string and pushes the drill cuttings up the annulus to the surface, where the cuttings are removed from the drilling fluid (drilling mud) by the shale shakers.

In a completed well, there may be many annuli. The 'A' annulus is the void between the production tubing and the smallest casing string. The A annulus can serve a number of crucial tasks, including gas lift and well kills. A normal well will also have a 'B' and frequently a 'C' annulus, between the different casing strings. These annuli do not normally have any connection to well bore fluids, but maintaining pressure in them is important in order to ensure integrity of the casing strings.

Though all annuli in a completed well are expected to be isolated from the production tubing and each other, connections allowing the flow of fluids between them may sometimes occur, due to either intervention or wear and tear. In these situations, it is said that there is "communication" between them.

During coiled tubing interventions, the void between the coil and the production tubing can also be considered an annulus and be used for circulation.

Blowout Preventer

A blowout preventer is a large, specialized valve used to seal, control and monitor oil and gas wells. Blowout preventers were developed to cope with extreme erratic pressures and uncontrolled flow (formation kick) emanating from a well reservoir during drilling. Kicks can lead to a potentially catastrophic event known as a blowout. In addition to controlling the downhole (occurring in the drilled hole) pressure and the flow of oil and gas, blowout preventers are intended to prevent tubing (e.g. drill pipe and well casing), tools and drilling fluid from being blown out of the wellbore (also known as bore hole, the hole leading to the reservoir) when a blowout threatens. Blowout preventers are critical to the safety of crew, rig (the equipment system used to drill a wellbore) and environment, and to the monitoring and maintenance of well integrity; thus blowout preventers are intended to be fail-safe devices.

The term BOP (an initialism rather than a spoken acronym, i.e., pronounced B-O-P, not "bop") is used in oilfield vernacular to refer to blowout preventers. The abbreviated term preventer, usually prefaced by a type (e.g. ram preventer), is used to refer to a single blowout preventer unit. A blowout preventer may also simply be referred to by its type (e.g. ram).

The terms blowout preventer, blowout preventer stack and blowout preventer system are commonly used interchangeably and in a general manner to describe an assembly of several stacked blowout preventers of varying type and function, as well as auxiliary components. A typical subsea deepwater blowout preventer system includes components such as electrical and hydraulic lines, control pods, hydraulic accumulators, test valve, kill and choke lines and valves, riser joint, hydraulic connectors, and a support frame. Two categories of blowout preventer are most prevalent: *ram* and *annular*. BOP stacks frequently utilize both types, typically with at least one annular BOP stacked above several ram BOPs. (A related valve, called an *inside blowout preventer*, *internal blowout preventer*, or *IBOP*, is positioned within, and restricts flow up, the drillpipe. This article does not address inside blowout preventer use.)

Blowout preventers are used at land and offshore rigs, and subsea. Land and subsea BOPs are secured to the top of the wellbore, known as the wellhead. BOPs on offshore rigs are mounted below the rig deck. Subsea BOPs are connected to the offshore rig above by a drilling riser that provides a continuous pathway for the drill string and fluids emanating from the wellbore. In effect, a riser extends the wellbore to the rig.

Use

The invention of blowout preventers was instrumental in reducing the incidence of oil gushers, blowouts, which are dangerous and costly. It has been reported that the preventers in use as of 2011 are not sufficient to prevent the most powerful of ruptures, such as experienced in the Deepwater Horizon blowout, indicating that substantial improvement is needed.

Blowout preventers come in a variety of styles, sizes and pressure ratings. Several individual units serving various functions are combined to compose a blowout preventer stack. Multiple blowout preventers of the same type are frequently provided for redundancy, an important factor in the effectiveness of fail-safe devices.

The primary functions of a blowout preventer system are to:

- Confine well fluid to the wellbore;
- Provide means to add fluid to the wellbore;
- Allow controlled volumes of fluid to be withdrawn from the wellbore.

Additionally, and in performing those primary functions, blowout preventer systems are used to:

- Regulate and monitor wellbore pressure;
- Centre and hang off the drill string in the wellbore;
- Shut in the well (e.g. seal the void, annulus, between drillpipe and casing);
- "Kill" the well (prevent the flow of formation fluid, influx, from the reservoir into the wellbore) ;
- Seal the wellhead (close off the wellbore);
- Sever the casing or drill pipe (in case of emergencies).

In drilling a typical high-pressure well, drill strings are routed through a blowout preventer stack toward the reservoir of oil and gas. As the well is drilled, drilling fluid, "mud", is fed through the drill string down to the drill bit, "blade", and returns up the wellbore in the ring-shaped void, annulus, between the outside of the drill pipe and the casing (piping that lines the wellbore). The column of drilling mud exerts downward hydrostatic pressure to counter opposing pressure from the formation being drilled, allowing drilling to proceed.

When a kick (influx of formation fluid) occurs, rig operators or automatic systems close the blowout preventer units, sealing the annulus to stop the flow of fluids out of the wellbore. Denser mud is then circulated into the wellbore down the drill string, up the annulus and out through the choke line at the base of the BOP stack through chokes (flow restrictors) until downhole pressure is overcome. Once "kill weight" mud extends from the bottom of the well to the top, the well has been "killed". If the integrity of the well is intact drilling may be resumed. Alternatively, if circulation is not feasible it may be possible to kill the well by "bullheading", forcibly pumping, in the heavier mud from the top through the kill line connection at the base of the stack. This is less desirable because of the higher surface pressures likely needed and the fact that much of the mud originally in the annulus must be forced into receptive formations in the open hole section beneath the deepest casing shoe.

If the blowout preventers and mud do not restrict the upward pressures of a kick, a blowout results, potentially shooting tubing, oil and gas up the wellbore, damaging the rig, and leaving well integrity in question. Since BOPs are important for the safety of the crew and natural environment, as well as the drilling rig and the wellbore itself, authorities recommend, and regulations require, that BOPs be regularly inspected, tested and refurbished. Tests vary from daily test of functions on critical wells to monthly or less frequent testing on wells with low likelihood of control problems.

Exploitable reservoirs of oil and gas are increasingly rare and remote, leading to increased subsea deepwater well exploration and requiring BOPs to remain submerged for as long as a year in extreme conditions. As a result, BOP assemblies have grown larger and heavier (e.g. a single ram-type BOP unit can weigh in excess of 30,000 pounds), while the space allotted for BOP stacks on existing offshore rigs has not grown commensurately. Thus a key focus in the technological development of BOPs over the last two decades has been limiting their footprint and weight while simultaneously increasing safe operating capacity.

Types

BOPs come in two basic types, *ram* and *annular*. Both are often used together in drilling rig BOP stacks, typically with at least one annular BOP capping a stack of several ram BOPs.

Ram Blowout Preventer

A ram-type BOP is similar in operation to a gate valve, but uses a pair of opposing steel plungers, rams. The rams extend toward the centre of the wellbore to restrict flow or retract open in order to permit flow. The inner and top faces of the rams are fitted with packers (elastomeric seals) that press against each other, against the wellbore, and around tubing running through the wellbore. Outlets at the sides of the BOP housing (body) are used for connection to choke and kill lines or valves.

Rams, or ram blocks, are of four common types: *pipe*, *blind*, *shear*, and *blind shear*.

Pipe rams close around a drill pipe, restricting flow in the annulus (ring-shaped space between concentric objects) between the outside of the drill pipe and the wellbore, but do not obstruct flow within the drill pipe. Variable-bore pipe rams can accommodate tubing in a wider range of outside diameters than standard pipe rams, but typically with some loss of pressure capacity and longevity.

Blind rams (also known as sealing rams), which have no openings for tubing, can close off the well when the well does not contain a drill string or other tubing, and seal it.

Shear rams cut through the drill string or casing with hardened steel shears.

Blind shear rams (also known as shear seal rams, or sealing shear rams) are intended to seal a wellbore, even when the bore is occupied by a drill string, by cutting through the drill string as the rams close off the well. The upper portion of the severed drill string is freed from the ram, while the lower portion may be crimped and the "fish tail" captured to hang the drill string off the BOP.

In addition to the standard ram functions, variable-bore pipe rams are frequently used as test rams in a modified blowout preventer device known as a stack test valve. Stack test valves are positioned at the bottom of a BOP stack and resist downward pressure (unlike BOPs, which resist upward pressures). By closing the test ram and a BOP ram about the drillstring and pressurizing the annulus, the BOP is pressure-tested for proper function.

The original ram BOPs of the 1920s were simple and rugged manual devices with minimal parts. The BOP housing (body) had a vertical well bore and horizontal ram cavity (ram guide chamber). Opposing rams (plungers) in the ram cavity translated horizontally, actuated by threaded ram shafts (piston rods) in the manner of a screw jack. Torque from turning the ram shafts by wrench or hand wheel was converted to linear motion and the rams, coupled to the inner ends of the ram shafts, opened and closed the well bore. Such screw jack type operation provided enough mechanical advantage for rams to overcome downhole pressures and seal the wellbore annulus.

Hydraulic rams BOPs were in use by the 1940s. Hydraulically actuated blowout preventers had many potential advantages. The pressure could be equalized in the opposing hydraulic cylinders causing the rams to operate in unison. Relatively rapid actuation and remote control were facilitated, and hydraulic rams were well-suited to high pressure wells. Because BOPs are fail-safe devices, efforts to minimize the complexity of the devices are still employed to ensure ram BOP reliability and longevity. As a result, despite the ever-increasing demands placed on them, state of the art ram BOPs are conceptually the same as the first effective models, and resemble those units in many ways.

Ram BOPs for use in deepwater applications universally employ hydraulic actuation. Threaded shafts are often still incorporated into hydraulic ram BOPs as lock rods that hold the ram in position after hydraulic actuation. By using a mechanical ram locking mechanism, constant hydraulic pressure need not be maintained. Lock rods may be coupled to ram shafts or not, depending on manufacturer. Other types of ram locks, such as wedge locks, are also used.

Typical ram actuator assemblies (operator systems) are secured to the BOP housing by removable bonnets. Unbolting the bonnets from the housing allows BOP maintenance and facilitates the substitution of rams. In that way, for example, a pipe ram BOP can be converted to a blind shear ram BOP.

Shear-type ram BOPs require the greatest closing force in order to cut through tubing occupying the wellbore. Boosters (auxiliary hydraulic actuators) are frequently mounted to the outer ends of a BOP's hydraulic actuators to provide additional shearing force for shear rams.

Ram BOPs are typically designed so that well pressure will help maintain the rams in their closed, sealing position. That is achieved by allowing fluid to pass to pass through a channel in the ram and exert pressure at the ram's rear and toward the centre of the wellbore. Providing a channel in the ram also limits the thrust required to overcome well bore pressure.

Single ram and double ram BOPs are commonly available. The names refer to the quantity of ram cavities (equivalent to the effective quantity of valves) contained in the unit. A double ram BOP is more compact and lighter than a stack of two single ram BOPs while providing the same functionality, and is thus desirable in many applications. Triple ram BOPs are also manufactured, but not as common.

Technological development of ram BOPs has been directed towards deeper and higher pressure wells, greater reliability, reduced maintenance, facilitated replacement of components, facilitated ROV intervention, reduced hydraulic fluid consumption, and improved connectors, packers, seals, locks and rams. In addition, limiting BOP weight and footprint are significant concerns to account for the limitations of existing rigs.

The highest-capacity large-bore ram blowout preventer on the market, as of July 2010, Cameron's EVO 20K BOP, has a hold-pressure rating of 20,000 psi, ram force in excess of 1,000,000 pounds, and a well bore diameter of 18.75 inches.

Annular Blowout Preventer

The annular blowout preventer was invented by Granville Sloan Knox in 1946; a U.S. patent for it was awarded in 1952. Often around the rig it is called the "Hydril", after the name of one of the manufacturers of such devices.

An annular-type blowout preventer can close around the drill string, casing or a non-cylindrical object, such as the kelly. Drill pipe including the larger-diameter tool joints (threaded connectors) can be "stripped" (i.e., moved vertically while pressure is contained below) through an annular preventer by careful control of the hydraulic closing pressure. Annular blowout preventers are also effective at maintaining a seal around the drillpipe even as it rotates during drilling. Regulations typically require that an annular preventer be able to completely close a wellbore, but annular preventers are generally not as effective as ram preventers in maintaining a seal on an open hole. Annular BOPs are typically located at the top of a BOP stack, with one or two annular preventers positioned above a series of several ram preventers.

An annular blowout preventer uses the principle of a wedge to shut in the wellbore. It has a donut-like rubber seal, known as an elastomeric packing unit, reinforced with steel ribs. The packing unit is situated in the BOP housing between the head and hydraulic piston. When the piston is actuated, its upward thrust forces the packing unit to constrict, like a sphincter, sealing the annulus or openhole. Annular preventers have only two moving parts, piston and packing unit, making them simple and easy to maintain relative to ram preventers.

The original type of annular blowout preventer uses a "wedge-faced" (conical-faced) piston. As the piston rises, vertical movement of the packing unit is restricted by the head and the sloped face of the piston squeezes the packing unit inward, toward the centre of the wellbore. In 1972, Ado N. Vujasinovic was awarded a patent for a variation on the annular preventer known as a spherical blowout preventer, so-named because of its spherical-faced head. As the piston rises the packing unit is thust upward against the curved head, which constricts the packing unit inward. Both types of annular preventer are in common use.

Control Methods

When rigs are drilled on land or in very shallow water where the wellhead is above the water line, BOPs are activated by hydraulic pressure from a remote accumulator. Several control stations will be mounted around the rig. They also can be closed manually by turning large wheel-like handles.

In deeper offshore operations with the wellhead just above the mudline on the sea floor, there are four primary ways by which a BOP can be controlled. The possible means are:

- Electrical Control Signal: sent from the surface through a control cable;
- Acoustical Control Signal: sent from the surface based on a modulated/encoded pulse of sound transmitted by an underwater transducer;
- ROV Intervention: remotely operated vehicles (ROVs) mechanically control valves and provide hydraulic pressure to the stack (via "hot stab" panels);
- Deadman Switch/Auto Shear: fail-safe activation of selected BOPs during an emergency, and if the control, power and hydraulic lines have been severed.

Two control pods are provided on the BOP for redundancy. Electrical signal control of the pods is primary. Acoustical, ROV intervention and dead-man controls are secondary.

An emergency disconnect system, or EDS, disconnects the rig from the well in case of an emergency. The EDS is also intended to automatically trigger the deadman switch, which closes the BOP, kill and choke valves. The EDS may be a subsystem of the BOP stack's control pods or separate. Pumps on the rig normally deliver pressure to the blowout preventer stack through hydraulic lines. Hydraulic accumulators are on the BOP stack enable closure of blowout preventers even if the BOP stack is disconnected from the rig. It is also possible to trigger the closing of BOPs automatically based on too high pressure or excessive flow.

Individual wells along the U.S. coastline may also be required to have BOPs with backup acoustic control. General requirements of other nations, including Brazil, were drawn to require this method. BOPs featuring this method may cost as much as US$500,000 more than those that omit the feature.

Deepwater Horizon Blowout

During the *Deepwater Horizon* drilling rig explosion incident on April 20, 2010, the blowout preventer should have been activated automatically, cutting the drillstring and sealing the well to preclude a blowout and subsequent oil spill in the Gulf of Mexico, but it failed to fully engage. Underwater robots (ROVs) later were used to manually trigger the blind shear ram preventer, to no avail.

As of May 2010 it is unknown why the blowout preventer failed. Chief surveyor John David Forsyth of the American Bureau of Shipping testified in hearings before the Joint Investigation of the Minerals Management Service and the U.S. Coast Guard investigating the causes of the explosion that his agency last inspected the rig's blowout preventer in 2005. BP representatives suggested that the preventer could have suffered a hydraulic leak. Gamma-ray imaging of the preventer conducted on May 12 and May 13, 2010 showed that the preventer's internal valves were partially closed and were restricting the flow of oil. Whether the valves closed automatically during the explosion or were shut manually by remotely operated vehicle work is unknown. A statement released by Congressman Bart Stupak revealed that, among other issues, the emergency disconnect system (EDS) did not function as intended and may have malfunctioned due to the explosion on the Deepwater Horizon.

The permit for the Macondo Prospect by the Minerals Management Service in 2009 did not require redundant acoustic control means. Inasmuch as the BOPs could not be closed successfully by underwater manipulation (ROV Intervention), pending results of a complete investigation, it is uncertain whether this omission was a factor in the blowout.

Documents discussed during congressional hearings June 17, 2010, suggested that a battery in the device's control pod was flat and that the rig's owner, Transocean, may have "modified" Cameron's equipment for the Macondo site (including incorrectly routing hydraulic pressure to a stack test valve instead of a pipe ram BOP) which increased the risk of BOP failure, in spite of warnings from their contractor to that effect. Another hypothesis is that a junction in the drilling pipe may have been positioned in the BOP stack in such way that its shear rams had an insurmountable thickness of material to cut through.

It was later discovered that a second piece of tubing got into the BOP stack at some point during the Macondo incident, potentially explaining the failure of the BOP shearing mechanism. As of July 2010 it is unknown whether the tubing might be casing that shot up through the well or perhaps broken drill pipe that dropped into the well.

On July 10, 2010 BP began operations to install a sealing cap, also known as a capping stack, atop the failed blowout preventer stack. Based on BP's video feeds of the operation the sealing cap assembly, called Top Hat 10, includes a stack of three blind shear ram BOPs manufactured by Hydril (a GE Oil & Gas company), one of Cameron's chief competitors. By July 15 the 3 ram capping stack had sealed the Macondo well, if only temporarily, for the first time in 87 days.

The U.S. government wants the failed blowout preventer to be replaced in case of any pressure that occurs when the relief well intersects with the well. On September 3 at 1:20 p.m. CDT the 300 ton failed blowout preventer was removed from the well and began being slowly lifted to the surface. Later that day a replacement blowout preventer was placed on the well. On September 4 at 6:54 p.m. CDT the failed blowout preventer reached the surface of the water and at 9:16 p.m. CDT it was placed in a special container on board the vessel Helix Q4000. The failed blowout preventer was taken to a NASA facility in Louisiana for examination by Det Norske Veritas (DNV).

On 20 March 2011, DNV presented their report to the US Department of Energy. Their primary conclusion was that the rams failed to shear through the oil pipe and seal it because it has buckled out of the line of action of the rams. They did not suggest any failure of actuation as would be caused by faulty batteries.

Bureau of Ocean Energy Management, Regulation and Enforcement

The Bureau of Ocean Energy Management, Regulation and Enforcement (BOEMRE), also known as the Bureau of Ocean Energy (BOE), and formerly known as the Minerals Management Service (MMS), is an agency of the United States Department of the Interior that manages the nation's natural gas, oil and other mineral resources on the outer continental shelf (OCS). The Offshore program, which manages the mineral resources on the OCS, is divided into three regions: Alaska, Gulf of Mexico, and the Pacific Ocean.

Headquartered in Washington, DC, the Agency receives most of its revenue from leasing federal lands and waters to oil and natural gas companies with a profit margin of 98%. It is among the top-five largest revenue sources to the federal government, the IRS being number-one. As the MMS (before transition to BOEMRE), the Agency's signature feature according to an informational trifold was that it had "become our Nation's leader in offshore energy development and the collection of royalties on behalf of the American Public." With respect to enforcement of regulations and safety, this same publication indicated that the "MMS also funds advanced scientific studies and enforces the highest safety and environmental standards." The Agency's mission statement was put more formally in its 2010 Budget Proposal:

MMS's mission is to manage the energy and mineral resources on the Outer Continental Shelf and Federal and American Indian mineral revenues to enhance public and trust benefits, promote responsible use, and realize fair value.

History

The Minerals Management Service was created on January 19, 1982. In January 1983, Congress passed the Federal Oil & Gas Royalty Management Act with the stated purpose:

To ensure that all oil and gas originated on the public lands and on the Outer Continental Shelf are properly accounted for under the direction of the Secretary of the Interior, and for other purposes.

The Secretary of the Interior at the time, James G. Watt, designated MMS as the administrative agency responsible for execution of activities under the Act.

With the passage of the Energy Policy Act of 2005, MMS was given authority to develop renewable energy projects such as wave, wind and current energy on the Outer Continental Shelf. As of 2010, the Agency was composed of two operating units, the MRM and OEMM.

- Offshore Energy and Minerals Management (OEMM) — Under the guidance of the 1953 Outer Continental Shelf Lands Act, the OEMM managed energy and mineral development in over 1.71 billion offshore acres of the Outer Continental Shelf (OCS) and annually disburses to the U.S. Treasury an average of $13 billion in minerals revenue.
- Minerals Revenue Management (MRM) — Through the MRM program, the Agency collected, accounted for, and disbursed mineral revenues from Federal and American Indian leases.

The Agency's offshore renewable energy program included development of renewable energy such as wind, wave, and solar.

Since its inception in 1982 through FY2008, the Agency had disbursed approximately $200 billion to Federal, state, and American Indian accounts. On June 21, 2010, the Minerals Management Service was renamed the Bureau of Ocean Energy Management, Regulation and Enforcement and reorganized.

Operations

As of 2009, the Agency employed about 1,600 people, which was proposed to grow by less than one hundred in 2010.

Organization

The BOEMRE was reorganized in May 2010 under the direction of Secretary of the Interior Ken Salazar following the Deepwater Horizon disaster. The bureau is organized into these three newly created agencies:

- Bureau of Ocean Energy Management — Responsible for leasing areas of the Outer Continental Shelf for conventional and renewable energy resources.
- Bureau of Safety and Environmental Enforcement — Responsible for ensuring comprehensive oversight, safety, and environmental protection in all offshore energy activities.
- Office of Natural Resources Revenue — Responsible for royalty and revenue management, including collection and distribution of revenue, auditing and compliance, and asset management.

Criticism and Controversies

Since the inception of the MMS, and in particular since the 1990s, the Agency has been embroiled or implicated in numerous scandals. For example, in 1990 MMS employees were linked to prostitution, and in 2008 the Department of Interior's Inspector General reported that MMS employees had participated in drug use and sexual activity with employees from the very energy firms they were to be regulating.

Collection of Oil and Gas Royalties

From the 1950s to at least 2002, drilling for oil and gas on federal lands and waters produced the second largest source of revenue for the federal government other than taxes. The Minerals Revenue Management (MRM) division of MSM was responsible for managing all royalties associated with both onshore and offshore oil and gas production from federal mineral leases. In 1997, in light of evidence that industry was getting around royalty regulations and underpaying royalties to the tune of billions of dollars, MMS proposed a more stringent rule to collect royalty payments in value (RIV), meaning in the form of cash payments from companies producing from federal leases. In response to that rule-making, industry proposed an alternative — "royalty-in-kind" (RIK) meaning in the form of actual oil or gas production. In fact, the industry opposed cash payments (RIV) and planned legal challenges to government efforts to establish regulations for fair market-based royalty payments. A pilot test of the RIK concept was conducted. The Bush administration allowed the pilot to expand to a full program, with industry support, even though the bill authorizing the program failed to pass in Congress. In FY2008, the RIK program accounted for more than 50% of the Agency's revenue collections. When MRM collected royalties-in-kind, the oil or gas received from producers was offered for sale by the U.S. Government on the open market and the proceeds from these sales were taken as revenues. The RIK program within MRM was responsible for managing these in-kind sales.

In 2003, the General Accounting Office (GAO) noted that the MMS had failed to develop "clear strategic objectives linked to statutory requirements nor collected the necessary information to effectively monitor and evaluate the Royalty-in-Kind Program". From 2003 to 2008, the GAO consistently challenged the legitimacy of the statistics published by the MMS that it used to support its claims that the RIK program was a success and justify its expansion.

Deficits in accounting practices, policies and procedures, and information systems used by the MMS led to concerns that the industry was significantly underpaying on their royalty obligations. Computer systems in use by the Agency were considered to be sufficiently inadequate that a failure to report revenue or provide RIK by an industry member could not be reliably detected. For instance, the GAO estimated that underpayments amounted to ~$160 million USD in 2006. The GAO also disputed the practice of tracking oil and gas RIK deliveries on a monthly rather than daily basis, a practice used by the MMS and supported by the Department of the Interior, but potentially prone to abuse by producers. Other contributing factors to the placing the accuracy of revenue streams at risk were insufficiently trained personnel and insufficient numbers of personnel working in the RIK program and a lack of standard reporting method by industry members, leading to manual rather than computer-based processing of more than half of the data required for RIK data inputs.

Citing its scandals and the persistent incapacity of the RIK program to fulfil its statutory obligations, Interior Secretary Salazar announced in September 2009 that the RIK program would be shut down. Due to existing lease contracts with RIK provisions, the program as of 2010 is still winding down. On October 7, 2009, the U.S. House Oversight Committee reported the loss of billions in revenue resulting from MMS mismanagement and cozy relationships with industry officials. According to Darrell Issa, the top Republican on the United States House Committee on Oversight and Government Reform, there may be a conflict of interest for the Minerals Management Service to collect revenue and also oversee safety.

Gifts, Gratuities and the Revolving Door

In September 2008, reports by the Inspector General of the Interior Department, Earl E. Devaney, were released that implicated over a dozen officials of the MMS of unethical and criminal conduct in the performance of their duties. The investigation found MMS employees had used cocaine and marijuana, and had sex with energy company representatives. MMS staff had also accepted gifts and free holidays amid "a culture of ethical failure", according to the investigation. The New York Times's summary states the investigation revealed "a dysfunctional organization that has been riddled with conflicts of interest, unprofessional behaviour and a free-for-all atmosphere for much of the Bush administration's watch."

A May 2010 inspector general investigation revealed that MMS regulators in the Gulf region had allowed industry officials to fill in their own inspection reports in pencil and then turned them over to the regulators, who traced over them in pen before submitting the reports to the agency. MMS staff had routinely accepted meals, tickets to sporting events, and gifts from oil companies. Staffers also used government computers to view pornography. In 2009 the regional supervisor of the Gulf region for MMS pled guilty and was sentenced to a year's probation in federal court for lying about receiving gifts from an offshore drilling contractor. "This deeply disturbing report is further evidence of the cozy relationship between MMS and the oil and gas industry," Salazar said.

The Project On Government Oversight (POGO) alleges that MMS has suffered from a systemic revolving door problem between the Department of Interior and the oil and gas industries. For example, thirteen months after departing as MMS director, Bush appointee Randall Luthi became president of the National Oceans Industries Association (NOIA) whose mission is to "to secure reliable access and a favourable regulatory and economic environment for the companies that develop the nation's valuable offshore energy resources in an environmentally responsible manner." Luthi succeeded Tom Fry, who was MMS director under the Clinton administration. Luthi and Fry represented precisely the industries their agency was tasked with being a watchdog over. Lower level administrators influencing MMS have also gone on to work for the companies they once regulated: In addition, Jimmy Mayberry served as Special Assistant to the Associate Director of Minerals Revenue Management (MRM), managed by MMS, from 2000 to January 2003. After he left, he created an energy consulting company that was awarded an MMS contract via a rigged bid. He was convicted along with a former MMS coworker Milton Dial who also came to work at the company. Both were found guilty of felony violation of conflict of interest law.

Deepwater Horizon Catastrophe and Restructuring

On May 11, 2010, in response to the Deepwater Horizon oil spill, Secretary of the Interior Ken Salazar announced that MMS would be restructured so that the safety and environmental functions are carried out by a unit with full independence from MMS in order to ensure that federal inspectors will have more tools, resources, and greater authority to enforce laws and regulations that apply to oil and gas companies operating on the Outer Continental Shelf. Another outcome of the spill was the retirement of the associate director for offshore energy and minerals management at the time of the spill, Chris Oynes.

MMS's regulatory decisions contributing to the 2010 oil spill included, in negligence, the decision that an acoustically-controlled shut-off valve (BOP) would not be required as a last resort against underwater spills at the site, MMS's failure to suggest other "fail-safe" mechanisms after a 2004 report raised questions about the reliability of the electrical remote-control devices., and the fact that MMS gave permission to BP and dozens of other oil companies to drill in the Gulf of Mexico without first getting required permits from the (National Oceanic and Atmospheric Administration that assess threats to endangered species and to assess the impact the drilling was likely to have on the gulf.

On May 19, 2010 Salazar announced that MMS will be broken up into three separate divisions, the Bureau of Ocean Energy Management, the Bureau of Safety and Environmental Enforcement, and the Office of Natural Resources Revenue, which will separately oversee energy leasing, safety enforcement, and revenue collection.

S. Elizabeth (Liz) Birnbaum served as the Director of the then named Minerals Management Service from July 15, 2009 to her resignation on May 27, 2010 amidst the Deepwater Horizon oil spill. On June 15, 2010 President Obama named Michael R. Bromwich, a former federal prosecutor and inspector general for the Justice Department, to head up efforts to restructure BOEMRE. Bob Abbey, then director of the Bureau of Land Management, took over as Acting Director of BOEMRE until his replacement could be confirmed. Amidst efforts to reorganize the bleagured agency, on June 21, 2010, Bromwich was sworn in as BOEMRE's new director, and Secretary of the Interior Ken Salazar issued a Secretarial Order that renamed the Minerals Management Service the Bureau of Ocean Energy Management, Regulation and Enforcement. Almost a year later, William K. Reilly, who co-chaired the commission charged with investigating the Horizon blowout, was quoted as saying "they changed the name, but all the people are the same" and "it's embarrassing" in reference to the current situation.

Chapter 9

Gas Oil Separation Plant

Gas Oil Separation Plant (GOSP) is a type of plant used primarily in the oil industry. The plant processes crude oil from the well head and separates gases and contaminants from the crude for reasons of safety, economy, and environmental protection. This makes the crude economically viable for storage, processing and export.

Produced crude oil is sent to a GOSP to be divided into oil and associated gases, which are then streamed to oil refineries and gas processing plants respectively. The products of these plants are then sent for supply and distribution.

Reasons for Processing

Pressure

The raw crude often leaves the well head under very high pressure. Production pressures of greater than 3000 pounds per square inch have been encountered in some fields. The high pressure makes transportation and storage difficult and dangerous.

Contaminants

Produced crude oil leaving the well head is both sour (contains hydrogen sulfide gas) and wet (contains water). The crude leaving the well head must be processed and treated to make it safe, environmentally acceptable, and economically viable for storage, processing and export.

Phase Separation

It is often appropriate to separate gases and liquids for separate use. This also involves the separation of oily and water liquid phases.

Gas Recovery

It is not appropriate to burn off the gases associated with crude oil. There are also economic reasons for processing and treating the produced crude. Recovering associated gases prevents wasting a natural resource, which was originally flared off.

Corrosion

There are also other economic reasons for GOSP. Removing contaminants from the crude, such as salt and hydrogen sulfide, protects the plant from corrosion damage. The initial processing of produced crude oil in a GOSP is also required to meet specifications of the crude for export and oil refining.

De-asphalter

A de-asphalter is a unit in a crude oil refinery or bitumen upgrader that separates asphalt from crude oil or bitumen.

The de-asphalter unit is usually placed after the vacuum distillation tower. It is usually a solvent de-asphalter unit, SDA. The SDA separates the asphalt from the feedstock because light hydrocarbons will dissolve aliphatic compounds but not asphaltenes. The output from the de-asphalter unit is de-asphalted oil ("DAO") and asphalt.

DAO from propane de-asphalting has the highest quality but lowest yield, whereas using pentane may double or triple the yield from a heavy feed, but at the expense of contamination by metals and carbon residues that shorten the life of downstream cracking catalysts. If the solvent is butane the unit will be referred to as a butane de-asphalter ("BDA") and if the solvent is propane, it will be called a propane de-asphalter ("PDA") unit.

Cold Heavy Oil Production with Sand

Cold heavy oil production with sand (CHOPS) is a technique for extracting difficult heavy crude oil where sand is used as a means enhancing the productivity of the oil well. CHOPS is now used as a "quasi primary" production approach in unconsolidated sandstones. Thousands of wells in Canada are now stably producing oil through CHOPS.

Unconventional Oil

Unconventional oil is petroleum produced or extracted using techniques other than the conventional (oil well) method. Oil industries and governments across the globe are investing in unconventional oil sources due to the increasing scarcity of conventional oil reserves. Although the depletion of such reserves is evident, unconventional oil production is a less efficient process and has greater environmental impacts than that of conventional oil production.

Burton Process

The Burton process is a thermal cracking process invented by William Merriam Burton and Robert Humphrey. The oil industry used it to double the production of gasoline in 1913. This thermal cracking process was patented on January 7, 1913 (Patent No. 1,049,667).

The first thermal cracking method, the Shukhov cracking process, was invented by Vladimir Shukhov (Patent of Russian empire No. 12926 on November 27, 1891).

In 1937 it was superseded by catalytic cracking.

It is still in use today to produce diesel.

Automated Tank Cleaning Machine

An automated tank cleaning machine is a machine used to clean cargo tanks such as those found in tank trucks, railroad cars, barges, and oil tankers. Genericized trademarks such as Butterworth machine and Gamajet are often used to refer to automatic tank cleaning machines regardless of their manufacturer.

Operation

Tanks must be cleaned from time to time for various reasons. One reason is to change the type of product carried inside a tank. Another is to allow the tank to be inspected or for maintenance to be performed within a tank.

Automated tank cleaning machines work in a manner similar to an irrigation sprinkler. Hot water forced through a jet nozzle rotates the nozzle. The nozzle's rotation moves the machine through a cleaning pattern. As the water sprays, the liquid is pumped out of the tank. Portable water washing systems are widely used, but tanks that are cleaned frequently may have a fixed system installed.

The Butterworth Type K machine is widely used. This model can clean a tank of up to 10,000,000 US gallons (38,000 m^3). It uses water with a pressure up to 250 pounds per square inch (1,700 kPa) and a temperature of up to 250 °F (121 °C). The water jet reaches up to 115 feet (35 m). Depending on the pressure used, a cleaning cycle can take from about 10 to 50 minutes and the machine uses between

15 US gallons (56.8 L; 12.5 imp gal) and 350 US gallons (1,324.9 L; 291.4 imp gal) per minute.

On most crude-oil tankers, a special crude oil washing system, or COW system, is part of the cleaning process. The COW system circulates hot crude oil through the fixed tank-cleaning system to remove wax and asphaltic deposits.

Although machines are often used to wash tanks, a final stage of manual cleaning known as mucking, is usually performed. Mucking requires protocols for entry into confined spaces and the use of airline respirators, protective clothing and safety observers.

Tank cleaning is dangerous in a number of ways. While tank barges can be cleaned in port, shipboard tanks are generally cleaned at sea. This is largely due to risks of fire and explosion inside the tanks.

History

The first automated tank cleaning machine was invented by Arthur Butterworth and patented in 1920. His goal was to limit the amount of time that workers had to spend inside tanks, and partially relieve them of a dangerous and laborious job. In 1925, Butterworth established a company to market the machine. Standard Oil New Jersey bought the company in 1930, and it later became a subsidiary of the Exxon Corporation.

In 1986, as part of an internal restructuring at Exxon, the Butterworth company was sold to Exxon management. Today the company is privately held, and headquartered in Houston, Texas.

Hydrodeoxygenation

Hydrodeoxygenation (HDO) is a hydrogenolysis process removing oxygenated compounds from fast pyrolysis or hydrothermal liquefaction derived bio-oil using commercial hydrotreating catalysts in the form of water.

Hydrotreating catalyst commonly used are nickel-molybdenum or cobalt-molybdenum on gamma alumina, zeolites (ZSM-5), palladium on carbon and platinum on carbon as well as alumina.

Infrared Open-path Detector

Infrared open-path gas detectors send out a beam of infrared light, detecting gas anywhere along the path of the beam. This linear 'sensor' is typically a few metres up to a few hundred metres in length. Open-path detectors can be contrasted with Infrared point sensors.

They are widely used in the petroleum and petrochemical industries, mostly to achieve very rapid gas leak detection for flammable gases at concentrations comparable to the lower flammable limit (typically a few percent by volume). They are also used, but so far to a lesser extent, in other industries where flammable concentrations can occur, such as in coal mining and water treatment. In principle the technique can also be used to detect toxic gases, for instance hydrogen sulfide, at the necessary parts-per-million concentrations, but the technical difficulties involved have so far prevented widespread adoption for toxic gases.

Usually, there are separate transmitter and receiver units at either end of a straight beam path. Alternatively, the source and receiver are combined, and the beam bounced off a retroreflector at the far end of the measurement path. For portable use, detectors have also been made which use the natural albedo of surrounding objects in place of the retroreflector. The presence of a chosen gas (or class of gases) is detected from its absorption of a suitable infrared wavelength in the beam. Rain, fog etc. in the measurement path can also reduce the strength of the received signal, so it is usual to make a simultaneous measurement at one or more reference wavelengths. The quantity of gas intercepted by the beam is then inferred from the ratio of the signal losses at the measurement and reference wavelengths. The calculation is typically carried out by a microprocessor which also carries out various checks to validate the measurement and prevent false alarms.

The measured quantity is the sum of all the gas along the path of the beam, sometimes termed the *path-integral concentration* of the gas. Thus the measurement has a natural bias (desirable in many applications) towards the total size of an unintentional gas release, rather than the concentration of the gas that has reached any particular point. Whereas the natural units of measurement for a Infrared point sensor are parts-per-million (ppm) or the percentage of the lower flammable limit (%LFL), the natural units of measurement for an open path detector are *ppm.metres* (ppmm) or *LFL.metres* (LFLm). For instance, the fire and gas safety system on an offshore platform in the North Sea typically has detectors set to a full-scale reading of 5LFLm, with low and high alarms triggered at 1LFLm and 3LFLm respectively.

Advantages and Disadvantages versus Fixed-point Detectors

An open path detector usually costs more than a single point detector, so there is little incentive for applications that play to a point detector's strengths: where the point detector can be placed at the known location of the highest gas concentration, and a relatively slow response is acceptable. The open path detector excels in outdoor situations where, even if the likely source of the gas release is known, the evolution of the developing cloud or plume is unpredictable. Gas will almost certainly enter a extended linear beam before finding its way to any single chosen point. Also, point detectors in exposed outdoor locations require weather shields to be fitted, increasing the response time significantly. Open path detectors can also show a cost advantage in any application where a row of point detectors would be required to achieve the same coverage, for instance monitoring along a pipeline, or around the perimeter of a plant. Not only will one detector replace several, but the costs of installation, maintenance, cabling etc. are likely to be lower.

An important consideration for both types is a realistic assessment of their availability in service, distinguishing carefully between revealed and unrevealed downtime. The latter, where a detector appears to be working but is insensitive to gas, is especially serious. For instance, point detectors of the catalytic type are prone to poisoning by silicones and H2S, or by clogging of the gauze or sinter by water or ice. Point detectors of the infrared type are immune from the former but not the latter mechanism. Open path detectors suffer downtime from anything that blocks the path of the beam, such as people, vehicles or thick fog.

However, they approach the ideal of eliminating unrevealed downtime because the loss of signal strength is easily made to generate a 'beam block' signal, distinct from zero gas or a hardware fault. Infrared detectors of both types, unlike passive sensors, invariably incorporate a microprocessor capable of self-checking the circuitry, so unrevealed downtime due to hardware failures is largely eliminated. The possibility of (revealed) downtime of an open path detector due to fog can be minimised by limiting the beam path to moderate lengths. In fog-prone areas, such as the North Sea, paths are often limited to 15 to 20 metres for this reason, even though the detectors are capable of much greater distances in clear air.

Component Parts

In principle any source of infrared radiation could be used, together with an optical system of lenses or mirrors to form the transmitted beam. In practice the following sources have been used, always with some form of modulation to aid the signal processing at the receiver:

- An incandescent light bulb, modulated by pulsing the current powering the filament or by a mechanical chopper. For systems used outdoors, it is difficult for an incandescent source to compete with the intensity of sunlight when the sun shines directly into the receiver. Also, it is difficult to achieve modulation frequencies distinguishable from those that can be produced naturally, for instance by heat shimmer or by sunlight reflecting off waves at sea.

- A gas-discharge lamp is capable of exceeding the spectral power of direct sunlight in the infrared, especially when pulsed. Modern open path systems typically use a xenon flashtube powered by a capacitor discharge. Such pulsed sources are inherently modulated.
- A semiconductor laser provides a relatively weak source, but one that can be modulated at high frequency in wavelength as well as amplitude. This property permits various signal processing schemes based on Fourier analysis, of use when the absorption of the gas is weak but narrow in spectral linewidth.

The precise wavelength passbands used must be isolated from the broad infrared spectrum. In principle any conventional spectrometer technique is possible, but the NDIR technique with multilayer dielectric filters and beamsplitters is most often used. These wavelength-defining components are usually located in the receiver, although one design has shared the task with the transmitter.

At the receiver, the infrared signal strengths are measured by some form of infrared detector. Generally photodiode detectors are preferred, and are essential for the higher modulation frequencies, whereas slower photoconductive detectors may be required for longer wavelength regions. The signals are fed to low-noise amplifiers, then invariably subject to some form of digital signal processing. The absorption coefficient of the gas will vary across the passband, so the simple Beer–Lambert law cannot be applied directly. For this reason the processing usually employs a calibration table, applicable for a particular gas, type of gas, or gas mixture, and sometimes configurable by the user.

Operating Wavelengths

The choice of infrared wavelengths used for the measurement largely defines the detector's suitability for a particular applications. Not only must the target gas (or gases) have a suitable absorption spectrum, the wavelengths must lie within a spectral window so the air in the beam path is itself transparent. These wavelength regions have been used:

- 3.4 μm region. All hydrocarbons and their derivatives absorb strongly, due to the C-H stretch mode of molecular vibration. It is commonly used in infrared point detectors where path lengths are necessarily short, and for open-path detectors requiring parts-per-million sensitivity. A disadvantage for many applications is that methane absorbs relatively weakly compared to heavier hydrocarbons, leading to large inconsistencies of calibration. For open-path detection of flammable concentrations the absorption for non-methane hydrocarbons is so strong that the measurement saturates, a significant gas cloud appearing 'black'. This wavelength region is beyond the transmission range of borosilicate glass, so windows and lenses must be made of more expensive materials and tend to be small in aperture.
- 2.3 μm region. All hydrocarbons and their derivatives have absorption coefficients appropriate for open path detection at flammable concentrations. A useful advantage in practical applications is that the detector's response to many different gases and vapours is relatively uniform when expressed in terms of the lower flammable limit. Borosilicate glass retains useful transmission in this wavelength region, allowing large aperture optics to be produced at moderate cost.
- 1.6 μm region. A wide range of gases absorb in the near-infrared. Typically the absorption coefficients are relatively weak, but light molecules show narrow, individually resolved spectral lines rather than broad bands. This results in relatively large values of the gradient and curvature of the absorption with respect to wavelength, enabling semiconductor laser-based systems to distinguish gas molecules very specifically; for instance hydrogen sulfide, or methane to the exclusion of heavier hydrocarbons.

History

The first open-path detector offered for routine industrial use, as distinct from research instruments built in small numbers, was the Wright and Wright 'Pathwatch' in the US, 1983. Acquired by Det-tronics in 1992, the detector operated in the 3.4 μm region with a powerful incandescent source and a mechanical chopper. It did not achieved large volume sales, mainly because of cost and doubts about long-term reliability with moving parts.

Beginning in 1985, Shell Research in UK was funded by Shell Natural Gas to develop an open-path detector with no moving parts.

The advantages of the 2.3 μm wavelength were identified, and a research prototype was demonstrated. This design had a combined transmitter-receiver with a corner-cube retroreflector at 50 m. It used a pulsed incandescent lamp, PbS photoconductive detectors in the gas and reference channels, and an Intel 8031 microprocessor for signal processing.

In 1987 Shell licenced this technology to Sieger-Zellweger (later Honeywell) who designed and marketed their industrial version as the 'Searchline', using a retro-reflective panel made up of multiple corner-cubes. This was the first open-path detector to be certified for use in hazardous areas and to have no moving parts. Later work by Shell Research used two alternately pulsed incandescent sources in the transmitter and a single PbS detectors in the receiver, avoiding zero drifts caused by the variable responsivity of PbS detectors. This technology was offered to Sieger-Zellweger, and later licensed to PLMS. a company part-owned by Shell Ventures UK. The PLMS GD4001/2 in 1991 were the first detectors to achieve a truly stable zero without moving parts or software compensation of slow drifts. They were also the first infrared gas detectors of any kind to be certified intrinsically safe. The Israeli company Spectronix (also Spectrex) made an important advance in 1996 with their SafEye, the first to use a flash tube source, followed by Sieger-Zellweger with their Searchline Excel in 1998. In 2001 the PLMS Pulsar, soon afterwards acquired by Dräger as their Polytron Pulsar, was the first detector to incorporate sensing to monitor the mutual alignment of the transmitter and receiver during both installation and routine operation.

Marine Loading Arm

A marine loading arm, also known as a mechanical loading arm, loading arm, or MLA is a device consisting of articulated steel pipes that connect a tankship such as an oil tanker or chemical tanker to a cargo terminal. Genericized trademarks such as Chiksan (often misspelled Chicksan) are often used to refer to marine loading arms regardless of their manufacturer.

Operation and Design

A marine loading arm is an alternative to direct hose hookups that is particularly useful for larger vessels and transfers at higher loading rates and pressures. Controlled manually or hydraulically, a loading arm employs swivel joints and can, to some extent, follow the movement of a moored vessel. Many loading arm systems feature quick-connect fittings. Gasket or o-ring arrangements are required to make a secure seal to the ship's manifold flange. A loading arm must be drained or closed off before the connection is broken off. This is usually done in two ways. For fuels such as gas oil and diesel, the lines can be blown out with high pressure air. In the case of fuels such as kerosene or petrol, the lines can be stripped with pumps.

Loading arms can handle both liquids and gases, in a wide range of viscosities and temperatures. Cargoes from liquid sulphur to liquefied natural gas are moved through marine loading arms. Loading arms service vessels in a wide range of sizes, from small river barges to the largest supertankers.

Various designs exist, and specific installations can be tailored for a given port based on considerations such as vessel size, cargo flow rate and cargo temperature. Environmental constraints, such as the range of tide, wind conditions, and earthquake tolerance can also affect choice of loading arm. A loading arm installation may include add-ons such as hydraulic or manual quick connect couplers, position monitoring systems, emergency release systems, and piggyback vapour return lines.

Compared to cargo hoses, the loading arm's main drawback is its comparative lack of flexibility.

History

Since the earliest days of tankships, the need to safely and efficiently transfer bulk liquid to a moored ship has been fundamental. An insufficient solution to this problem led to one of the world's first oil tanker disasters. In 1881, the Branobel tanker *Nordenskjöld* was taking on kerosene in Baku. The ship was connected to the pier with a simple piece of pipe. While loading, the ship was hit by a gust of wind and the cargo pipe carrying was jerked away from the hold. Kerosene then spilled onto the deck and down into the engineroom, where mechanics were working in the light of kerosene lanterns. The ship then exploded, killing half the crew. Ludvig Nobel responded to the disaster by creating a flexible, leakproof loading pipe which was much more resistant to spills.

Chiksan brand marine loading arm manufacturer FMC Technologies claims to have built the world's first all-steel marine loading arm in 1956 and to have over 8,000 units installed worldwide.

Single Buoy Mooring

Single Point Mooring (SPM) is a loading buoy anchored offshore that serves as a mooring point and interconnect for tankers loading or offloading gas or fluid products. SPMs are the link between geostatic subsea manifold connections and weathervaning tankers. They are capable of handling any size ship, even very large crude carriers (VLCC) where no alternative facility is available.

Parts of a Single Point Mooring

There are four parts in the total buoying system: the body of the buoy, mooring and anchoring elements, product transfer system and other components. All these parts are equally important.

Buoy Body

The buoy body usually is supported on static legs attached to the seabed, with a rotating part above water level connected to the (off)loading tanker. The two sections are linked by a roller bearing, referred to as the "main bearing". The moored tanker can freely weathervane around the buoy and find a stable position due to this arrangement.

Mooring and Anchoring Parts

Moorings fix the buoy to the seabed. Buoy design must account for the behaviour of the buoy given applicable wind, wave and current conditions and tanker sizes. This determines the optimum mooring arrangement and size of the various mooring leg components. Anchoring points are greatly dependent on local soil condition.

Mooring Components

- Anchors or piles - To connect the mooring to the seabed.
- Anchor Chain
- Chainstoppers - To connect the chains to the buoy.
- Manpower.

Hawser Arrangement

A tanker is moored to a buoy by means of a hawser arrangement. Oil Companies International Marine Forum (OCIMF) standards are available for mooring systems.

The hawser arrangement usually consists of nylon rope, which is shackled to an integrated mooring uni-joint on the buoy deck. At the tanker end of the hawser, a chafe chain is connected to prevent damage from the tanker fairlead. A load pin can be applied to the mooring uni-joint on the buoy deck to measure hawser loads.

Hawser systems use either one or two ropes depending on the largest size of vessel which would be moored to the buoy. The ropes would either be single leg or grommet leg type ropes. These are usually connected to an OCIMF chafe chain on the export tanker side (either type A or B depending on the maximum size of the tanker and the mooring loads). This chafe chain would then be held in the chain stopper on board the export tanker. A basic hawser system would consist of the following (working from the buoy outwards):

Buoy Side Shackle and bridle assembly for connection to the padeye on the buoy; Mooring Hawser Shackle; Mooring Hawser; Chafe Chain Assembly; Support Buoy; Pick-up/Messenger Lines; Marker Buoy for retrieval from the water.

Under OCIMF recommendations, the hawser arrangement would normally be purchased as a full assembly from a manufacturer.

Product Transfer System

The heart of each buoy is the product transfer system. From a geostatic location, e.g. a pipeline end manifold (PLEM) located on the seabed, this system transfers products to the offtake tanker.

The basic product transfer system components are:

- Flexible Subsea Hoses, - generally referred to as "Risers".
- Floating Hose String(s)
- Product Swivel, Valves and Piping.

Risers

The risers are flexible hoses that connect the subsea piping to the buoy. Configuration of these risers can vary depending on water-depth, sea state, buoy motions, etc.

Floating Hose String

Floating hose string(s) connect the buoy to the offloading tanker. The hose string can be equipped with a breakaway coupling to prevent rupture of hoses/hawser and subsequent oil spills.

Product Swivel

The product swivel is the connection between the geostatic and the rotating parts of the buoy. The swivel enables an offloading tanker to rotate with respect to the mooring buoy. Product swivels range in size depending on the size of attached piping and risers. Product swivels can provide one or several independent paths for fluids, gases, electrical signals or power. Swivels are equipped with a multiple seal arrangement to minimise the possibility of leakage of product into the environment.

Other Components

Other possible components of SPMs are:

- A Boat Landing, providing access to the buoy deck,
- Fendering to protect the buoy,
- Lifting and Handling Equipment to aid materials handling,
- Navigational Aids for maritime visibility,
- An Electrical Subsystem to enable valve operation and to power navigation aids or other equipment.

Torpedo (Petroleum)

A torpedo is an explosive device used, especially in the early days of the petroleum industry, to fracture the surrounding rock at the bottom of an oil well to stimulate the flow of oil and to remove built-up paraffin wax that would restrict the flow. Earlier torpedoes used gunpowder, but the use of nitroglycerin eventually became widespread. The development of hydraulic fracturing rendered torpedoes obsolete, and is the primary fracturing process used today.

Use

A torpedo consisted of canisters that were filled with an explosive and lowered into a well via a rope or wire. Gunpowder was used in the first torpedoes, but nitroglycerin was found to work better despite its instability. The well is usually filled with water to prevent the explosion from escaping upwards. Originally, the topmost canister had a percussion cap that was to detonate the main charge. An iron weight was dropped down the well to set the torpedo off. After incidents of premature explosions, a second method was developed in which a tube of the explosive was placed in a larger tube that was packed with sand. A fuse was wound around the inner tube, connected to a blasting cap. When the torpedo was to be used, the inner tube was filled with nitroglycerin and corked; the fuse was lit and torpedo was dropped down the well.

Torpedoes were generally used to remove buildup of paraffin wax from an oil well. Before the use of torpedoes caught on, boiling water or benzene was often poured down wells to try to dissolve the paraffin. Torpedoes were also used to fracture the rock to allow the oil to flow more easily.

History

Edward A. L. Roberts developed the first torpedo and submitted a patent application in November 1864. Roberts, an American Civil War veteran, came up with the concept of using water to "tamp" the resulting explosion, after watching Confederate artillery rounds explode in a canal at the Battle of Fredericksburg. Roberts developed his first torpedoes in 1865 and 1866.

In November 1866 he was granted a patent on his torpedo application, and founded the Roberts Petroleum Torpedo Company. William Reed also developed a torpedo design and went on to found a rival company "for the purpose of infringing and breaking down the Roberts patent.

Roberts charged $100-200 per torpedo as well as a royalty amounting to D $_{15}$ of the increased oil production. To avoid paying the exorbitant fees, an owner of a well would often hire men who illegally produced their own torpedoes and used them at night—the practice giving rise to term "moonlighting". Roberts spent $250,000 to protect his patent from the "moonlighters" by hiring the Pinkerton National Detective Agency and filing numerous lawsuits. Roberts' torpedo patents expired in 1879.

Torpedoes manufactured today use modern explosives, with the last nitroglycerin torpedo being used on May 5, 1990.

Thermal Depolymerization

Thermal depolymerization (TDP) is a depolymerization process using hydrous pyrolysis for the reduction of complex organic materials (usually waste products of various sorts, often biomass and plastic) into light crude oil. It mimics the natural geological processes thought to be involved in the production of fossil fuels. Under pressure and heat, long chain polymers of hydrogen, oxygen, and carbon decompose into short-chain petroleum hydrocarbons with a maximum length of around 18 carbons.

Similar Processes

Thermal depolymerisation is similar to other processes which use superheated water as a major step to produce fuels, such as direct Hydrothermal Liquefaction. These are distinct from processes using dry materials to depolymerize, such as pyrolysis. The term Thermochemical Conversion (TCC) has also been used for conversion of biomass to oils, using superheated water, although it is more usually applied to fuel production via pyrolysis. Other commercial scale processes include the "SlurryCarb" process operated by EnerTech, which uses similar technology to decarboxylate wet solid biowaste, which can then be physically dewatered and used as a solid fuel called

E-Fuel. The plant at Rialto is said to be able to process 683 tons of waste per day. The Hydro Thermal Upgrading (HTU) process uses superheated water to produce oil from domestic waste. A demonstration plant is due to start up in The Netherlands said to be capable of processing 64 tons of biomass (dry basis) per day into oil. Thermal depolymerisation differs in that it contains a hydrous process followed by an anhydrous cracking/distillation process.

History

Thermal depolymerization is similar to the geological processes that produced the fossil fuels used today, except that the technological process occurs in a timeframe measured in hours. Until recently, the human-designed processes were not efficient enough to serve as a practical source of fuel—more energy was required than was produced.

In U. S. patent 2,177,557, issued in 1939, Bergstrom and Cederquist discuss a method for obtaining oil from wood in which the wood is heated under pressure in water with a significant amount of calcium hydroxide added to the mixture. In the early 1970s Herbert R. Appell and coworkers worked with hydrous pyrolysis methods, as exemplified by U. S. patent 3,733,255 (issued in 1973), which discusses the production of oil from sewer sludge and municipal refuse by heating the material in water, under pressure, and in the presence of carbon monoxide.

An approach that exceeded break-even was developed by Illinois microbiologist Paul Baskis in the 1980s and refined over the next 15 years. The technology was finally developed for commercial use in 1996 by Changing World Technologies (CWT). Brian S. Appel (CEO of CWT) took the technology in 2001 and expanded and changed it into what is now referred to as TCP (Thermal Conversion Process), and has applied for several patents. A Thermal Depolymerization demonstration plant was completed in 1999 in Philadelphia by Thermal Depolymerization, LLC, and the first full-scale commercial plant was constructed in Carthage, Missouri, about 100 yards (91m) from ConAgra Foods' massive Butterball turkey plant, where it is expected to process about 200 tons of turkey waste into 500 barrels (79 m^3) of oil per day.

Theory and Process

In the method used by CWT, the water improves the heating process and contributes hydrogen to the reactions.

In the Changing World Technologies (CWT) process, the feedstock material is first ground into small chunks, and mixed with water if it is especially dry. It is then fed into a pressure vessel reaction chamber where it is heated at constant volume to around 250 °C.

Similar to a pressure cooker (except at much higher pressure), steam naturally raises the pressure to 600 psi (4 MPa) (near the point of saturated water). These conditions are held for approximately 15 minutes to fully heat the mixture, after which the pressure is rapidly released to boil off most of the water.

The result is a mix of crude hydrocarbons and solid minerals. The minerals are removed, and the hydrocarbons are sent to a second-stage reactor where they are heated to 500 °C, further breaking down the longer hydrocarbon chains. The hydrocarbons are then sorted by fractional distillation, in a process similar to conventional oil refining.

The CWT company claims that 15 to 20% of feedstock energy is used to provide energy for the plant. The remaining energy is available in the converted product. Working with turkey offal as the feedstock, the process proved to have yield efficiencies of approximately 85%; in other words, the energy contained in the end products of the process is 85% of the energy contained in the inputs to the process (most notably the energy content of the feedstock, but also including electricity for pumps and natural gas or woodgas for heating).

If one considers the energy content of the feedstock to be free (i.e., waste material from some other process), then 85 units of energy are made available for every 15 units of energy consumed in process heat and electricity. This means the "Energy Returned on Energy Invested" (EROEI) is (6.67), which is comparable to other energy harvesting processes. Higher efficiencies may be possible with drier and more carbon-rich feedstocks, such as waste plastic.

By comparison, the current processes used to produce ethanol and biodiesel from agricultural sources have EROEI in the 4.2 range, when the energy used to produce the feedstocks is accounted for (in this case, usually sugar cane, corn, soybeans and the like).

These EROEI values are not directly comparable, because these EROEI calculations include the energy cost to produce the feedstock, whereas the above EROEI calculation for thermal depolymerization process (TDP) does not.

The process breaks down almost all materials that are fed into it. TDP even efficiently breaks down many types of hazardous materials, such as poisons and difficult-to-destroy biological agents such as prions.

Feedstocks and Outputs with Thermal Depolymerization

Figure: *Average TDP Feedstock Outputs*

Feedstock	***Oils***	***Gases***	***Solids (mostly carbon based)***	***Water (Steam)***
Plastic bottles	70%	16%	6%	8%
Medical waste	65%	10%	5%	20%
Tires	44%	10%	42%	4%
Turkey offal	39%	6%	5%	50%
Sewage sludge	26%	9%	8%	57%
Paper (cellulose)	8%	48%	24%	20%

(Note: Paper/cellulose contains at least 1% minerals, which was probably grouped under carbon solids.)

Carthage Plant Products

As reported on 04/02/2006 by Discover Magazine, the Carthage plant was producing 500 barrels per day (79 m^3/d) of oil made from 270 tons of turkey entrails and 20 tons of hog lard. This represents an oil yield of 22.3 percent. The Carthage, Missouri plant produces API 40+, a high value crude oil. It contains light and heavy naphthas, a kerosene, and a gas oil fraction, with essentially no heavy fuel oils, tars, asphaltenes or waxes. It can be further refined to produce No. 2 and No. 4 fuel oils.

Figure: *TDP-40 Oil Classification by D-5443 PONA method*

Output Material	***% by Weight***
Paraffins	22%
Olefins	14%
Naphthenes	3%
Aromatics	6%
C14/C14+	55%
	100%

The fixed carbon solids produced by the TDP process have multiple uses as a filter, a fuel source and a fertilizer. It can be used as activated carbon in wastewater treatment, as a fertilizer, or as a fuel similar to coal.

Advantages

The process can break down organic poisons, due to breaking chemical bonds and destroying the molecular shape needed for the poison's activity. It is likely to be highly effective at killing pathogens, including prions. It can also safely remove heavy metals from the samples by converting them from their ionized or organometallic forms to their stable oxides which can be safely separated from the other products. Along with similar processes, it is a method of recycling the energy content of organic materials without first removing the water. It can produce liquid fuel, which separates from the water physically without need for drying. Other methods to recover energy often require pre-drying (e.g. burning, pyrolysis) or produce gaseous products (e.g. anaerobic digestion).

Potential Sources of Waste Inputs

The United States Environmental Protection Agency estimates that in 2006 there were 251 million tons of municipal solid waste, or 4.6 pounds generated per day per person in the USA. Much of this mass is considered unsuitable for oil conversion.

Limitations

The process only breaks long molecular chains into shorter ones, so small molecules such as carbon dioxide or methane cannot be converted to oil through this process. However, the methane in the feedstock is recovered and burned to heat the water that is an essential part of the process. In addition, the gas can be burned in a combined heat and power plant, consisting of a gas turbine which drives a generator to create electricity, and a heat exchanger to heat the process input water from the exhaust gas. The electricity can be sold to the power grid, for example under a Feed-in Tariff scheme. This also increases the overall efficiency of the process (already said to be over 85% of feedstock energy content).

Another option is to sell the methane product as biogas. For example, biogas can be compressed, much like natural gas, and used to power motor vehicles. Many agricultural and animal wastes could be processed, but many of these are already used as fertilizer, animal feed, and, in some cases, as feedstocks for paper mills or as boiler fuel. Energy crops constitute another potentially large feedstock for thermal depolymerization.

Current Status

Reports in 2004 claimed that the facility was selling products at 10% below the price of equivalent oil, but its production costs were low enough that the plant produced a profit. At the time it was paying for turkey waste.

The plant then consumed 270 tons of turkey offal (the full output of the turkey processing plant) and 20 tons of egg production waste daily. In February 2005, the Carthage plant was producing about 400 barrels per day (64 m^3/d) of crude oil.

In April 2005 the plant was reported to be running at a loss. Further 2005 reports summarized some economic setbacks which the Carthage plant encountered since its planning stages. It was thought that concern over mad cow disease would prevent the use of turkey waste and other animal products as cattle feed, and thus this waste would be free.

As it turned out, turkey waste may still be used as feed in the United States, so that the facility must purchase that feed stock at a cost of $30 to $40 per ton, adding $15 to $20 per barrel to the cost of the oil. Final cost, as of January 2005, was $80/barrel ($1.90/gal).

The above cost of production also excludes the operating cost of the thermal oxidizer and scrubber added in May 2005 in response to odor complaints.

A biofuel tax credit of roughly $1 per US gallon (26 ¢/L) on production costs was not available because the oil produced did not meet the definition of "biodiesel" according to the relevant American tax legislation. The Energy Policy Act of 2005 specifically added thermal depolymerization to a $1 renewable diesel credit, which became effective at the end of 2005, allowing a profit of $4/barrel of output oil.

Company Expansion

The company has explored expansion in California, Pennsylvania, and Virginia, and is presently examining projects in Europe, where animal products cannot be used as cattle feed. TDP is also being considered as an alternative means for sewage treatment in the United States.

Smell Complaints

The pilot plant in Carthage was temporarily shut down due to smell complaints. It was soon restarted when it was discovered that few of the odors were generated by the plant. Furthermore, the plant agreed to install an enhanced thermal oxidizer and to upgrade its air scrubber system under a court order. Since the plant is located only four blocks from the tourist-attracting town centre, this has strained relations with the mayor and citizens of Carthage.

According to a company spokeswoman, the plant has received complaints even on days when it is not operating. She also contended that the odors may not have been produced by their facility, which is located near several other agricultural processing plants.

On December 29, 2005, the plant was ordered by the state governor to shut down once again over allegations of foul odours as reported by MSNBC.

As of March 7, 2006, the plant has begun limited test runs to validate it has resolved the odor issue.

As of August 24, 2006, the last lawsuit connected with the odor issue has been dismissed and the problem is acknowledged as fixed. In late November, however, another complaint was filed over bad smells. This complaint was closed on January 11 of 2007 with no fines assessed.

Status as of February 2009

A May 2003 article in Discover magazine stated, "Appel has lined up federal grant money to help build demonstration plants to process chicken offal and manure in Alabama and crop residuals and grease in Nevada. Also in the works are plants to process turkey waste and manure in Colorado and pork and cheese waste in Italy. He says the first generation of depolymerization centers will be up and running in 2005. By then it should be clear whether the technology is as miraculous as its backers claim."

However, as of August 2008, the only operational plant listed at the company's website is the initial one in Carthage, Missouri.

Changing World Technology applied for an IPO on August 12; 2008, hoping to raise $100 million.

The unusual Dutch Auction type IPO failed possibly because CWT has lost nearly $20 million with very little revenue.

Similar Technologies

Plasma Converters use powerful electric arcs to reduce and extract energy from waste.

Oil Sands

Bituminous sands, colloquially known as oil sands or tar sands, are a type of unconventional petroleum deposit. The sands contain naturally occurring mixtures of sand, clay, water, and a dense and extremely viscous form of petroleum technically referred to as bitumen (or colloquially "tar" due to its similar appearance, odour, and colour). Oil sands are found in large amounts in many countries throughout the world, but are found in extremely large quantities in Canada and Venezuela.

The crude bitumen contained in the Canadian oil sands is described by Canadian authorities as "petroleum that exists in the semi-solid or solid phase in natural deposits. Bitumen is a thick, sticky form of crude oil, so heavy and viscous (thick) that it will not flow unless heated or diluted with lighter hydrocarbons. At room temperature, it is much like cold molasses". Venezuelan authorities often refer to similar types of crude oil as *extra-heavy oil*, because Venezuelan reservoirs are warmer and the oil is somewhat less viscous, allowing it to flow more easily.

Oil sands reserves have only recently been considered to be part of the world's oil reserves, as higher oil prices and new technology enable them to be profitably extracted and upgraded to usable products. They are often referred to as unconventional oil or crude bitumen, in order to distinguish the bitumen extracted from oil sands from the free-flowing hydrocarbon mixtures known as crude oil traditionally produced from oil wells.

Making liquid fuels from oil sands requires energy for steam injection and refining. This process generates two to four times the amount of greenhouse gases per barrel of final product as the "production" of conventional oil. If combustion of the final products is included, the so-called "Well to Wheels" approach, oil sands extraction, upgrade and use emits 10 to 45% more greenhouse gases than conventional crude.

History

The exploitation of bituminous deposits and seeps dates back to paleolithic times. The earliest known use of bitumen was by Neanderthals, some 40,000 years ago. Bitumen has been found adhering to stone tools used by Neanderthals at sites in Syria. After the arrival of Homo sapiens, humans used bitumen for construction of buildings and water proofing of reed boats, among other uses. In ancient Egypt, the use of bitumen was important in creating Egyptian mummies—in fact, the word mummy is derived from the Arab word *mûmiyyah*, which means bitumen.

In ancient times, bitumen was primarily a Mesopotamian commodity used by the Sumerians and Babylonians, although it was also found in the Levant and Persia. The area along the Tigris and Euphrates rivers was littered with hundreds of pure bitumen seepages. The Mesopotamians used the bitumen for waterproofing boats and buildings. In North America, the early European fur traders found Canadian First Nations using bitumen from the vast Athabasca oil sands to waterproof their birch bark canoes. In Europe, they were extensively mined near the European city of Pechelbronn, where the vapour separation process was in use in 1742.

The name *tar sands* was applied to bituminous sands in the late 19th and early 20th century. People who saw the bituminous sands during this period were familiar with the large amounts of tar residue produced in urban areas as a by-product of the manufacture of coal gas for urban heating and lighting.

The word "tar" to describe these natural bitumen deposits is really a misnomer, since, chemically speaking, tar is a man-made substance produced by the destructive distillation of organic material, usually coal. Since then, coal gas has almost completely been replaced by natural gas as a fuel, and coal tar as a material for paving roads has been replaced by the petroleum product asphalt. Naturally occurring bitumen is chemically more similar to asphalt than to tar, and the term *oil sands* (or oilsands) is more commonly used in the producing areas than *tar sands* because synthetic oil is what is manufactured from the bitumen.

Oil sands are now considered a serious alternative to conventional crude oil, since crude oil is becoming scarcer. Oil sands and oil shale have the potential to generate oil for centuries.

Reserves

Many countries in the world have large deposits of oil sands, including the United States, Russia, and various countries in the Middle East. However, the world's largest deposits occur in two countries: Canada and Venezuela, each of which have oil sand reserves approximately equal to the world's total reserves of conventional crude oil. As a result of the development of Canadian oil sands reserves, 44% of Canadian oil production in 2007 was from oil sands, with an additional 18% being heavy crude oil, while light oil and condensate had declined to 38% of the total.

Because growth of oil sands production has exceeded declines in conventional crude oil production, Canada has become the largest supplier of oil and refined products to the United States, ahead of Saudi Arabia and Mexico. Venezuelan production is also very large, but due to political problems within its national oil company, estimates of its production data are not reliable. Outside analysts believe Venezuela's oil production has declined in recent years, though there is much debate on whether this decline is depletion-related or not.

In October 2009, the USGS updated the Orinoco oil sands (Venezuela) mean estimated recoverable value to 513 billion barrels (81.6×10^9 m^3), making it "one of the world's largest recoverable" oil deposits.

Production

Bituminous sands are a major source of unconventional oil. Conventional crude oil is normally extracted from the ground by drilling oil wells into a petroleum reservoir, allowing oil to flow into them under natural reservoir pressures, although artificial lift and techniques such as water flooding and gas injection are usually required to maintain production as reservoir pressure drops toward the end of a field's life. Because extra-heavy oil and bitumen flow very slowly, if at all, toward producing wells under normal reservoir conditions, the sands must be extracted by strip mining or the oil made to flow into wells by in situ techniques, which reduce the viscosity by injecting steam, solvents, and/or hot air into the sands. These processes can use more water and require larger amounts of energy than conventional oil extraction, although many conventional oil fields also require large amounts of water and energy to achieve good rates of production.

This is because heavy crude feedstock needs pre-processing before it is fit for conventional refineries. This pre-processing is called 'upgrading', the key components of which are as follows:

1. removal of water, sand, physical waste, and lighter products
2. catalytic purification by hydrodemetallisation (HDM), hydrodesulfurization (HDS) and hydrodenitrogenation (HDN)
3. hydrogenation through carbon rejection or catalytic hydrocracking (HCR)

As carbon rejection is very inefficient and wasteful in most cases, catalytic hydrocracking is preferred in most cases. All these processes take large amounts of energy and water, while emitting more carbon dioxide than conventional oil.

Catalytic purification and hydrocracking are together known as hydroprocessing. The big challenge in hydroprocessing is to deal with the impurities found in heavy crude, as they poison the catalysts over time. Many efforts have been made to deal with this to ensure high activity and long life of a catalyst. Catalyst materials and pore size distributions are key parameters that need to be optimized to deal with these challenge and this varies from place to place, depending on the kind of feedstock present.

At the present time, only Canada has a large-scale commercial oil sands industry, though a small amount of oil from oil sands is produced in Venezuela. Because of increasing oil sands production, Canada has become the largest single supplier of oil and products to the United States. Oil sands now are the source of almost half of Canada's oil production, although due to the 2008 economic downturn work on new projects has been deferred, while Venezuelan production has been declining in recent years. Oil is not produced from oil sands on a significant level in other countries.

Transportation and Refining

The heavy crude oil or crude bitumen extracted from oil sands is a viscous, solid or semisolid form that does not easily flow at normal oil pipeline temperatures, making it difficult to transport to market and expensive to process into gasoline, diesel fuel, and other products. Despite the difficulty and cost, oil sands are now being mined by energy companies on a vast scale to extract the bitumen, which is then converted into synthetic oil (syncrude) by bitumen upgraders, or refined directly into petroleum products by specialized refineries.

As Oil Source, by Location

Canada

Canada is the largest supplier of crude oil and refined products to the United States, supplying about 20% of total U.S. imports, and exports more oil and products to the U.S. than it consumes itself. In 2006, bitumen production averaged 1.25 million barrels per day (200,000 m^3/d) through 81 oil sands projects, representing 47% of total Canadian petroleum production. This proportion is expected to increase in coming decades as bitumen production grows while conventional oil production declines.

Most of the oil sands of Canada are located in three major deposits in northern Alberta. These are the Athabasca-Wabiskaw oil sands of north northeastern Alberta, the Cold Lake deposits of east northeastern Alberta, and the Peace River deposits of northwestern Alberta. Between them they cover over 140,000 square kilometres (54,000 sq mi)—an area larger than England—and hold proven reserves of 1.75 trillion barrels (280×10^9 m^3) of bitumen in place. About 10% of this, or 173 billion barrels (27.5×10^9 m^3), is estimated by the government of Alberta to be recoverable at current prices, using current technology, which amounts to 97% of Canadian oil reserves and 75% of total North American petroleum reserves. The Cold Lake deposits extend across the Alberta's eastern border into Saskatchewan. In addition to the Alberta oil sands, there are major oil sands deposits on Melville Island in the Canadian Arctic islands, which are unlikely to see commercial production in the foreseeable future.

The Alberta oil sand deposits contain at least 85% of the world's reserves of natural bitumen (representing 40% of the combined crude bitumen and extra-heavy crude oil reserves in the world), but are the only bitumen deposits concentrated enough to be economically recoverable for conversion to synthetic crude oil at current prices. The largest bitumen deposit, containing about 80% of the Alberta total, and the only one suitable for surface mining, is the Athabasca Oil Sands along the Athabasca River. The mineable area (as defined by the Alberta government) includes 37 townships covering about 3,400 square kilometres (1,300 sq mi) near Fort McMurray. The smaller Cold Lake deposits are important because some of the oil is fluid enough to be extracted by conventional methods. All three Alberta areas are suitable for production using in-situ methods, such as cyclic steam stimulation (CSS) and steam assisted gravity drainage (SAGD).

The Alberta oil sands have been in commercial production since the original Great Canadian Oil Sands (now Suncor) mine began operation in 1967. A second mine, operated by the Syncrude consortium, began operation in 1978 and is the biggest mine of any type in the world. The third mine in the Athabasca Oil Sands, the Albian Sands consortium of Shell Canada, Chevron Corporation, and Western Oil Sands Inc. [purchased by Marathon Oil Corporation in 2007] began operation in 2003. Petro-Canada was also developing a $33 billion Fort Hills Project, in partnership with UTS Energy Corporation and Teck Cominco, which lost momentum after the 2009 merger of Petro-Canada into Suncor. If approved, upgraders were slated to begin output in 4–5 years.

With the development of new in-situ production techniques such as steam assisted gravity drainage, and with the oil price increases since 2003, there were several dozen companies planning nearly 100 oil sands projects in Canada, totaling nearly $100 billion in capital investment. With 2007 crude oil prices significantly in excess of the current average cost of production of $28 per barrel of bitumen, all of these projects appear likely to be profitable. However, bitumen production costs are rising rapidly, with production cost increases of 55% since 2005, due to shortages of labour and materials.

The minority Conservative government of Canada, pressured to do more on the environment, announced in its 2007 budget that it would phase out some oil sands tax incentives over coming years. The provision allowing accelerated write-off of oil sands investments will be phased out gradually, so projects that had relied on them can proceed. For new projects the provision will be phased out between 2011 and 2015.

With oil prices setting new highs in 2007, tax incentives were no longer necessary to encourage oil sands projects in Canada. In July of that year, Royal Dutch Shell released its 2006 annual report and announced that its Canadian oil sands unit made an after tax profit of $21.75 per barrel, nearly double its worldwide profit of $12.41 per barrel on conventional crude oil. A few days later, Shell announced it had filed for regulatory approval to build a $27 billion oil sands refinery in Alberta, one of $38 billion in new oil sands projects announced that week.

Oil sands development in Alberta is strongly opposed by some Canadian and other environmentalists. A pipeline from Alberta to Gulf coast refineries, Keystone XL is under consideration.

Venezuela

Located in eastern Venezuela, north of the Orinoco River, the Orinoco oil belt vies with the Canadian oil sand for largest known accumulation of bitumen in the world. Venezuela prefers to call its oil sands "extra heavy oil", and although the distinction is somewhat academic, the extra heavy crude oil deposit of the Orinoco Belt represent nearly 90% of the known global reserves of extra heavy crude oil, and nearly 45% of the combined crude bitumen and extra-heavy crude oil reserves in the world.

Bitumen and extra-heavy oil are closely related types of petroleum, differing only in the degree by which they have been degraded from the original crude oil by bacteria and erosion. The Venezuelan deposits are less degraded than the Canadian deposits and are at a higher temperature (over 50 degrees Celsius versus freezing for northern Canada), making them easier to extract by conventional techniques.

Although Venezuela's extra-heavy oil is easier to produce than Canada's bitumen, it is still too heavy to transport by pipeline or process in normal refineries. Lacking access to first-world capital and technological prowess, Venezuela has not been able to design and build the kind of upgraders and heavy oil refineries that Canada has. In the early 1980s, the state oil company, PDVSA, developed a method of using the extra-heavy oil resources by emulsifying it with water (70% extra-heavy oil, 30% water) to allow it to flow in pipelines. The resulting product, called Orimulsion, can be burned in boilers as a replacement for coal and heavy fuel oil with only minor modifications. Unfortunately, the fuel's high sulfur content and emission of particulates make it difficult to meet increasingly strict international environmental regulations.

Further development of the Venezuelan resources has been impeded by political unrest. Venezuela is less politically stable than a country such as Canada, and a two-month strike in 2002–2003 by employees of the state oil company was followed by the dismissal of nearly 20,000 staff. As tensions resolved, strike leaders pointed to the reduction in Venezuela's domestic crude output as an argument that Venezuela's oil production had fallen. However, Venezuela's oil sands crude production, which sometimes wasn't counted in its total, has increased from 125,000 bbl/d (19,900 m^3/d) to 500,000 bbl/d (79,000 m^3/d) between 2001 and 2006 (Venezuela's figures; IAEA says 300,000 bbl/d).

USA

In the United States, oil sands resources are primarily concentrated in Eastern Utah. With a total of 32 billion barrels (5.1×10^9 m^3) of oil (known and potential) in eight major deposits in the Utah counties of Carbon, Garfield, Grand, Uintah, and Wayne. Currently, oil is not produced from oil sands on a significant commercial level in the United States, although the U.S. imports twenty percent of its oil and refined products from Canada, and over fifty percent of Canadian oil production is from oil sands. In addition to being much smaller than the oil sands deposits in Alberta, Canada, the U.S. oil sands are hydrocarbon wet, whereas the Canadian oil sands are water wet. As a result of this difference, extraction techniques for the Utah oil sands will be different than those used for the Alberta oil sands. A considerable amount of research has been done in the quest for commercially viable production technology to be employed in the development of the Utah oil sands. A special concern is the relatively arid climate of eastern Utah, as a large amount of water may be required by some processing techniques. Section 526 of the *Energy Independence And Security Act* prohibits United States government agencies from buying oil produced by processes that produce more greenhouse gas emissions than would traditional petroleum including oil sands.

Other Countries

Several other countries hold oil sands deposits which are smaller by orders of magnitude. Russia holds oil sands in two main regions.

The Volga-Urals basins (in and around Tatarstan), which is an important but very mature province in terms of conventional oil, holds large amounts of oil sands in a shallow permian formation. Exploitation has not gone beyond pilot stage yet. Other, less known, deposits are located in eastern Siberia.

In the Republic of the Congo, the Italian oil company Eni have announced in May 2008 a project to develop the small oil sands deposit in order to produce 40,000 barrels per day (6,400 m^3/d) in 2014. Reserves are estimated between 0.5 and 2.5 billion barrels (400×10^6 m^3).

In Madagascar, Tsimiroro and Bemolanga are two heavy oil sands deposits with a pilot well already producing small amounts of oil in Tsimiroro. and larger scale exploitation in the early planning phase.

Extraction Process

Surface Mining

Since Great Canadian Oil Sands (now Suncor) started operation of its mine in 1967, bitumen has been extracted on a commercial scale from the Athabasca Oil Sands by surface mining. In the Athabasca sands there are very large amounts of bitumen covered by little overburden, making surface mining the most efficient method of extracting it. The overburden consists of water-laden muskeg (peat bog) over top of clay and barren sand. The oil sands themselves are typically 40 to 60 metres deep, sitting on top of flat limestone rock. Originally, the sands were mined with draglines and bucket-wheel excavators and moved to the processing plants by conveyor belts. In recent years companies such as Syncrude and Suncor have switched to much cheaper shovel-and-truck operations using the biggest power shovels (100 or more tons) and dump trucks (400 tons) in the world. This has held production costs to around $27 per barrel of synthetic crude oil despite rising energy and labour costs.

After excavation, hot water and caustic soda (NaOH) is added to the sand, and the resulting slurry is piped to the extraction plant where it is agitated and the oil skimmed from the top. Provided that the water chemistry is appropriate to allow bitumen to separate from sand and clay, the combination of hot water and agitation releases bitumen from the oil sand, and allows small air bubbles to attach to the bitumen droplets. The bitumen froth floats to the top of separation vessels, and is further treated to remove residual water and fine solids. Bitumen is much thicker than traditional crude oil, so it must be either mixed with lighter petroleum (either liquid or gas) or chemically split before it can be transported by pipeline for upgrading into synthetic crude oil.

The bitumen is then transported and eventually upgraded into synthetic crude oil. About two tons of oil sands are required to produce one barrel (roughly 1/8 of a ton) of oil. Originally, roughly 75% of the bitumen was recovered from the sand. However, recent enhancements to this method include Tailings Oil Recovery (TOR) units which recover oil from the tailings, Diluent Recovery Units to recover naptha from the froth, Inclined Plate Settlers (IPS) and disc centrifuges. These allow the extraction plants to recover well over 90% of the bitumen in the sand. After oil extraction, the spent sand and other materials are then returned to the mine, which is eventually reclaimed.

Alberta Taciuk Process technology extracts bitumen from oil sands through a dry-retorting. During this process, oil sand is moved through a rotating drum, cracking the bitumen with heat and producing lighter hydrocarbons. Although tested, this technology is not in commercial use yet.

Four oil sands mines are currently in operation and two more (Jackpine and Kearl) are in the initial stages of development. The original Suncor mine opened in 1967, while the Syncrude mine started in 1978, Shell Canada opened its Muskeg River mine (Albian Sands) in 2003 and Canadian Natural Resources Ltd opened its Horizon Project in 2009. New mines under construction or undergoing approval include Shell Canada's, Imperial Oil's Kearl Oil Sands Project, Synenco Energy's, and Suncor's.

It is estimated that approximately 90% of the Alberta oil sands and nearly all of Venezuelan sands are too far below the surface to use open-pit mining. Several in-situ techniques have been developed to extract this oil.

Cold Flow

In this technique, also known as cold heavy oil production with sand (CHOPS), the oil is simply pumped out of the sands, often using progressive cavity pumps. This only works well in areas where the oil is fluid enough. It is commonly used in Venezuela (where the extra-heavy oil is at 50 degrees Celsius), and also in the Wabasca, Alberta Oil Sands, the southern part of the Cold Lake, Alberta Oil Sands and the Peace River Oil Sands. It has the advantage of being cheap and the disadvantage that it recovers only 5-6% of the oil in place.

Some years ago Canadian oil companies discovered that if they removed the sand filters from the wells and produced as much sand as possible with the oil, production rates improved remarkably. This technique became known as Cold Heavy Oil Production with Sand

(CHOPS). Further research disclosed that pumping out sand opened "wormholes" in the sand formation which allowed more oil to reach the wellbore. The advantage of this method is better production rates and recovery (around 10%) and the disadvantage that disposing of the produced sand is a problem. A novel way to do this was spreading it on rural roads, which rural governments liked because the oily sand reduced dust and the oil companies did their road maintenance for them. However, governments have become concerned about the large volume and composition of oil spread on roads, so in recent years disposing of oily sand in underground salt caverns has become more common.

Cyclic Steam Stimulation (CSS)

The use of steam injection to recover heavy oil has been in use in the oil fields of California since the 1950s. The Cyclic Steam Stimulation or "huff-and-puff" method has been in use by Imperial Oil at Cold Lake since 1985 and is also used by Canadian Natural Resources at Primrose and Wolf Lake and by Shell Canada at Peace River. In this method, the well is put through cycles of steam injection, soak, and oil production. First, steam is injected into a well at a temperature of 300 to 340 degrees Celsius for a period of weeks to months; then, the well is allowed to sit for days to weeks to allow heat to soak into the formation; and, later, the hot oil is pumped out of the well for a period of weeks or months. Once the production rate falls off, the well is put through another cycle of injection, soak and production. This process is repeated until the cost of injecting steam becomes higher than the money made from producing oil. The CSS method has the advantage that recovery factors are around 20 to 25% and the disadvantage that the cost to inject steam is high.

Steam Assisted Gravity Drainage (SAGD)

Steam assisted gravity drainage was developed in the 1980s by the Alberta Oil Sands Technology and Research Authority and fortuitously coincided with improvements in directional drilling technology that made it quick and inexpensive to do by the mid 1990s. In SAGD, two horizontal wells are drilled in the oil sands, one at the bottom of the formation and another about 5 metres above it. These wells are typically drilled in groups off central pads and can extend for miles in all directions. In each well pair, steam is injected into the upper well, the heat melts the bitumen, which allows it to flow into the lower well, where it is pumped to the surface. SAGD has proved to be a major breakthrough in production technology since it is cheaper than

CSS, allows very high oil production rates, and recovers up to 60% of the oil in place. Because of its very favourable economics and applicability to a vast area of oil sands, this method alone quadrupled North American oil reserves and allowed Canada to move to second place in world oil reserves after Saudi Arabia. Most major Canadian oil companies now have SAGD projects in production or under construction in Alberta's oil sands areas and in Wyoming. Examples include Japan Canada Oil Sands Ltd's (JACOS) project, Suncor's Firebag project, Nexen's Long Lake project, Suncor's (formerly Petro-Canada's) MacKay River project, Husky Energy's Tucker Lake and Sunrise projects, Shell Canada's Peace River project, Cenovus Energy's Foster Creek and Christina Lake developments, ConocoPhillips' Surmont project, Devon Canada's Jackfish project, and Derek Oil & Gas's LAK Ranch project. Alberta's OSUM Corp has combined proven underground mining technology with SAGD to enable higher recovery rates by running wells underground from within the oil sands deposit, thus also reducing energy requirements compared to traditional SAGD. This particular technology application is in its testing phase.

Vapour Extraction Process (VAPEX)

VAPEX is similar to SAGD but instead of steam, hydrocarbon solvents are injected into the upper well to dilute the bitumen and allow it to flow into the lower well. It has the advantage of much better energy efficiency over steam injection, and it does some partial upgrading of bitumen to oil right in the formation. It is very new, but the process has attracted much attention from oil companies, who are beginning to experiment with it.

The above three methods are not mutually exclusive. It is becoming common for wells to be put through one CSS injection-soak-production cycle to condition the formation prior to going to SAGD production, and companies are experimenting with combining VAPEX with SAGD to improve recovery rates and lower energy costs.

Toe to Heel Air Injection (THAI)

This is a very new and experimental method that combines a vertical air injection well with a horizontal production well. The process ignites oil in the reservoir and creates a vertical wall of fire moving from the "toe" of the horizontal well toward the "heel", which burns the heavier oil components and upgrades some of the heavy bitumen into lighter oil right in the formation. Historically fireflood projects have not worked out well because of difficulty in controlling the flame front and a propensity to set the producing wells on fire. However, some oil companies feel the THAI method will be more controllable and practical, and have the advantage of not requiring energy to create steam. Advocates of this method of extraction state that it uses less freshwater, produces 50% less greenhouse gases, and has a smaller footprint than other production techniques.

Petrobank Energy and Resources Ltd. has reported encouraging results from their test wells in Alberta, with production rates of up to 400 barrels per day (64 m^3/d) per well, and the oil upgraded from 8 to 12 API degrees. The company hopes to get a further 7-degree upgrade from its CAPRI (controlled atmospheric pressure resin infusion) system, which pulls the oil through a catalyst lining the lower pipe.

Combustion Overhead Gravity Drainage (COGD)

This is an experimental method that employs a number of vertical air injection wells above a horizontal production well located at the base of the bitumen pay zone. An initial Steam Cycle similar to CSS is used to prepare the bitumen for ignition and mobility. Following that cycle, air is injected into the vertical wells, igniting the upper bitumen and mobilizing (through heating) the lower bitumen to flow into the production well. It is expected that COGD will result in water savings of 80% compared to SAGD.

Environmental Issues

Like all petroleum production, oil sands operations can have an adverse effect on the environment. Oil sands projects can potentially affect: the land when the bitumen is initially mined and with large deposits of toxic chemicals; the water during the separation process and through the drainage of rivers; and the air due to the release of carbon dioxide and other emissions. Additional indirect environmental effects are that the petroleum products produced are mostly burned, releasing carbon dioxide into the atmosphere. Heavy metals such as vanadium, nickel, lead, cobalt, mercury, chromium, cadmium, arsenic, selenium, copper, manganese, iron and zinc are present in oil sands.

Air

The Wood Buffalo Environmental Association (WBEA) monitors the air in the Regional Municipality of Wood Buffalo continuously. This is done through a variety of air, land and human monitoring programs. The information collected is openly shared with stakeholders and the public.

Since 1995, monitoring in the oil sands region shows improved or no change in long term air quality for the five key air quality pollutants

— carbon monoxide, nitrogen dioxide, ozone, fine particulate matter (PM2.5) and sulfur dioxide — used to calculate the Air Quality Index. Air monitoring has shown significant increases in exceedances of hydrogen sulfide (H_2S) both in the Fort McMurray area and near the oil sands upgraders.

Hydrogen sulfide is the chemical compound with the formula H_2S. This colorless, toxic and flammable gas is responsible for the foul odour of rotten eggs. Hydrogen sulfide gas occurs naturally in crude petroleum, natural gas, volcanic gases and hot springs. It also can result from bacterial breakdown of organic matter and be produced by human and animal wastes.

In 2007, the Alberta government issued an Environmental Protection Order to Suncor Energy Inc. The order comes in response to numerous occasions when ground level concentration (GLC) for H_2S exceeded acceptable standards. Environmental Protection Orders are issued under the authority of Alberta's Environmental Protection and Enhancement Act. Alberta Environment can issue Environmental Protection Orders to remedy environmental problems where there has been a release of a substance that has caused or may cause an adverse effect to the environment.

Land

A large part of oil sands mining operations involves clearing trees and brush from a site and removing the "overburden" — the topsoil, muskeg, sand, clay and gravel — that sits atop the oil sands deposit. Approximately two tons of oil sands are needed to produce one barrel of oil (roughly 1/8 of a ton). As a condition of licensing, projects are required to implement a reclamation plan. The mining industry asserts that the boreal forest will eventually colonize the reclaimed lands, but that their operations are massive and work on long-term timeframes. As of 2006/2007 (the most recent data available), about 420 km^2 (160 sq mi) of land in the oil sands region have been disturbed, and 65 km^2 (25 sq mi) of that land is under reclamation. In March 2008, Alberta issued the first-ever oil sands land reclamation certificate to Syncrude Canada Ltd. for the 1.04 km^2 (0.40 sq mi) parcel of land known as Gateway Hill approximately 35 km (22 mi) north of Fort McMurray. Several reclamation certificate applications for oil sands projects are expected within the next 10 years.

Water

Between 2 to 4.5 volume units of water are used to produce each volume unit of synthetic crude oil (SCO) in an ex-situ mining operation.

Despite recycling, almost all of it ends up in tailings ponds, which, as of 2007, covered an area of approximately 50 km^2 (19 sq mi). In SAGD operations, 90 to 95 percent of the water is recycled and only about 0.2 volume units of water is used per volume unit of bitumen produced. Large amounts of water are used for oil sands operations – Greenpeace gives the number as 349 million cubic metres per year, twice the amount of water used by the city of Calgary.

It is unclear if this is the amount of water they are licensed to remove from the Athabasca or the actual use and how up to date the statistic is. The Athabasca River is also much larger than Bow and Elbow rivers that flow through Calgary.

The Athabasca River is the ninth longest river in Canada running 1,231 km (765 mi) from the Athabasca Glacier in west-central Alberta to Lake Athabasca in northeastern Alberta. The average annual flow just downstream of Fort McMurray is 633 cubic metres per second with its highest daily average measuring 1200 cubic metres per second.

Current water license allocations totals about 1.8 percent of the Athabasca river flow. Actual use in 2006 was about 0.4 percent. In addition, the Alberta government sets strict limits on how much water oil sands companies can remove from the Athabasca River. According to the Water Management Framework for the Lower Athabasca River, during periods of low river flow water consumption from the Athabasca River is limited to 1.3 per cent of annual average flow. The province of Alberta is also looking into cooperative withdrawal agreements between oil sands operators.

In October 2009, Suncor Energy announced it was seeking government approval for a new process to recover tailings called Tailings Reduction Operations (TRO), which accelerates the settling of fine clay, sand, water, and residual bitumen in ponds after oil sands extraction.

The technology involves dredging mature tailings from a pond bottom, mixing the suspension with a polymer flocculent, and spreading the sludge-like mixture over a "beach" with a shallow grade. According to the company, the process could reduce the time for water reclamation from tailings to weeks rather than years, with the recovered water being recycled into the oil sands plant. In addition to reducing the number of tailing ponds, Suncor claims TRO could reduce the time to reclaim a tailing pond from 40 years at present to 7–10 years, with land rehabilitation continuously following 7 to 10 years behind the mining operations.

In December 2010, the Oil Sands Advisory Panel, commissioned by former environment minister Jim Prentice, found that the system in place for monitoring water quality in the region, including work by the Regional Aquatic Monitoring Program, the Alberta Water Research Institute, the Cumulative Environmental Management Association and others, was piecemeal and should become more comprehensive and coordinated.

A major hindrance to the monitoring of oil sands produced waters has been the lack of identification of individual compounds present. By better understanding the nature of the highly complex mixture of compounds, including naphthenic acids, it may be possible to monitor rivers for leachate and also to remove toxic components. Such identification of individual acids has for many years proved to be impossible but a recent breakthrough in analysis has begun to reveal what is in the oil sands produced waters.

Climate Change

The production of bitumen and synthetic crude oil emits more greenhouse gas (GHG) than the production of conventional crude oil, and has been identified as the largest contributor to GHG emissions growth in Canada, as it accounts for 40 million tons of CO_2 emissions per year. According to the Canadian Association of Petroleum Producers, Environment Canada claims the oil sands make up 5% of Canada's greenhouse gas emissions, or 0.1% of global greenhouse gas emissions. It predicts the oil sands will grow to make up 8% of Canada's greenhouse gas emissions by 2015. Environmentalists argue that the availability of more oil for the world made possible by oil sands production in itself raises global emissions of CO_2.

While the emissions per barrel of bitumen produced decreased 26% over the decade 1992–2002, total emissions were expected to increase due to higher production levels. As of 2006, to produce one barrel of oil from the oil sands released almost 75 kg (170 lb) of GHG with total emissions estimated to be 67 megatonnes (66,000,000 long tons; 74,000,000 short tons) per year by 2015.

In January 2008, the Alberta government released *Alberta's 2008 Climate Change Strategy*. Alberta's emissions are projected to grow to 400 megatonnes (Mt) by 2050, largely due to forecast growth in the oil sands sector. The new plan aims to cut the projected 400 Mt in half by 2050, with a 139 Mt reduction coming from carbon capture and storage — the bulk of those reductions (100 Mt) will come from activities related to oil sands production.

A federal court of Canada ruling on March 6, 2008, found the approval of Imperial Oil Ltd.'s $8-billion oil sands mine insufficient on climate change and greenhouse gas emissions. Proposals in the regulatory system at that date included mines by Total SA of France, by Anglo-Dutch Royal Dutch Shell and by Petro-Canada, as well as steam-injection projects by EnCana of Calgary.

A 2009 study by CERA estimated that production from Canada's oil sands emits "about 5 percent to 15 percent more carbon dioxide, over the "well-to-wheels" lifetime analysis of the fuel, than average crude oil." With coal's CO_2 emissions about one-third higher than conventional oil emissions, that makes the oil sands' emissions equal to about 90% of the CO_2 released from coal.

On September 21, 2010, a study by "IHS (Information Handling Services) Cambridge Energy Research Associates (IHS CERA)" found that fuels made from Canadian oil sands "result in significantly lower greenhouse gas (GHG) emissions than many commonly cited estimates... Oil sands products imported to the United States result in GHG emissions that are, on average, six percent higher than the average crude consumed in the country. This level places oil sands on par with other sources of U.S. crude imports, including crudes from Nigeria, Venezuela and some domestically produced oil, the report finds."

Carbon Dioxide Sequestration

To offset greenhouse gas emissions from the oil sands and elsewhere in Alberta, sequestering carbon dioxide emissions inside depleted oil and gas reservoirs has been proposed. This technology is inherited from Enhanced oil recovery methods, which have been in use for several decades. In July 2008, the Alberta government announced a C$2 billion fund to support sequestration projects in Alberta power plants (largely coal) and oil sands extraction and upgrading facilities.

Aquatic Life Deformities

There is conflicting research on the effects of the oil sands on aquatic life in the Canadian oil sands development. In 2007, Environment Canada completed a study that shows high deformity rates in fish embryos exposed to the oil sands. David W. Schindler, a limnologist from the University of Alberta, co-authored a study on Alberta's oil sands' contribution of aromatic polycyclic compounds, some of which are known carcinogens, to the Athabasca River and its tributaries. Scientists, local doctors, and residents supported a letter sent to the Prime Minister in September 2010 calling for an independent study of Lake Athabasca (which is downstream of the oil sands) to be initiated due to the rise of deformities and tumours found in fish caught there. The bulk of the research that defends the oil sands development is done by the Regional Aquatics Monitoring Program, RAMP. RAMP studies show that deformity rates are normal compared to historical data and the deformity rates in rivers upstream of the oil sands. It should be noted that RAMP is affiliated with the oil industry and its research data is submitted to environmental government agencies but unlike academia where peer review happens on a per study basis, RAMP does a peer review of the entire organization only once every five years.

Concerns of Environmentalists

The environmental impact caused by oil sand extraction is frequently criticized by environmental groups such as Greenpeace. Greenpeace is concerned with the social and health costs, pollution of the Athabasca River, air toxins, loss of farmland, removal of Boreal Forest and the growth of greenhouse gas emissions.

Oil sands extraction is generally held to be more environmentally damaging than conventional crude oil — carbon dioxide "well-to-pump" emissions, for example, are estimated to be about 1.3-1.7 times that of conventional crude.

Public Health Impacts

Concerns have been raised by a number of groups concerning the potential negative impacts that the tar sands have on public health, including claims of higher than normal rates of cancer among residents of Fort Chipewyan. In August, 2011, the Alberta government initiated a provincial health study of the link between the higher rates of cancer and the tar sands. It has also been suggested that other wildlife has been negatively affected by the oil sands; for instance, moose were found in a 2006 study to have as high as 453 times the acceptable levels of arsenic in their systems, though later studies lowered this to 17 to 33 times the acceptable level (still causing a danger for human consumption).

Input Energy

Approximately 1.0 – 1.25 gigajoule of energy is needed to extract a barrel of bitumen and upgrade it to synthetic crude. As of 2006, most of this is produced by burning natural gas. Since a barrel of oil equivalent is about 6.117 gigajoules, this extracts about 5 or 6 times as much energy as is consumed. Energy efficiency is expected to improve to 0.7 gigajoules of energy per barrel by 2015, giving an EROEI of about 9. However, since natural gas production in Alberta peaked in 2001 and has been static ever since, it is likely oil sands requirements will be met by cutting back natural gas exports to the U.S. Alternatives to natural gas exist and are available in the oil sands area. Bitumen can itself be used as the fuel, consuming about 30-35% of the raw bitumen per produced unit of synthetic crude. Nexen's Long Lake project will use a proprietary deasphalting technology to upgrade the bitumen, using asphaltene residue fed to a gasifier whose syngas will be used by a cogeneration turbine and a hydrogen producing unit, providing all the energy needs of the project: steam, hydrogen, and electricity.

Thus, it will produce syncrude without consuming natural gas, but the capital cost is very high.

Coal is widely available in Alberta and is inexpensive, but produces large amounts of greenhouse gases. Nuclear power is another option which has been proposed, but did not appear to be economic as of 2005.

In early 2007 the Canadian House of Commons Standing Committee on Industry, Science and Technology considered that the use of nuclear power to process oil sands could reduce CO_2 emissions and help Canada meet its Kyoto commitments, as it would require nearly 12 GW to meet production growth to 2015, but the implications of building reactors in northern Alberta were not yet well understood.

Energy Alberta Corporation announced in 2007 that they had applied for a license to build a new nuclear plant at Lac Cardinal, 30 km west of the town of Peace River. The application would see an initial twin AECL *Advanced CANDU Reactor* ACR-1000 plant go online in 2017, producing 2.2 GW (electric). At 6.117 GJ/barrel, this is equivalent to conserving 31,074 barrels per day (4,940.4 m^3/d). On November 30, 2007 Bruce Power, which operates eight CANDU reactors in Ontario, signed a letter of intent to acquire Energy Alberta and take over the project.

Steam-assisted Gravity Drainage

Steam Assisted Gravity Drainage (SAGD) is an enhanced oil recovery technology for producing heavy crude oil and bitumen. It is an advanced form of steam stimulation in which a pair of horizontal wells are drilled into the oil reservoir, one a few metres above the other. Low pressure steam is continuously injected into the upper wellbore to heat the oil and reduce its viscosity, causing the heated oil to drain into the lower wellbore, where it is pumped out.

Description

In the SAGD process, two parallel horizontal oil wells are drilled in the formation, one about 4 to 6 metres above the other. The upper well injects steam, possibly mixed with solvents, and the lower one collects the heated crude oil or bitumen that flows out of the formation, along with any water from the condensation of injected steam. The basis of the process is that the injected steam forms a "steam chamber" that grows vertically and horizontally in the formation. The heat from the steam reduces the viscosity of the heavy crude oil or bitumen which allows it to flow down into the lower wellbore. The steam and gases rise because of their low density compared to the heavy crude oil below, ensuring that steam is not produced at the lower production well.

The gases released, which include methane, carbon dioxide, and usually some hydrogen sulfide, tend to rise in the steam chamber, filling the void space left by the oil and, to a certain extent, forming an insulating heat blanket above the steam. Oil and water flow is by a countercurrent, gravity driven drainage into the lower well bore. The condensed water and crude oil or bitumen is recovered to the surface by pumps such as progressive cavity pumps that work well for moving high-viscosity fluids with suspended solids.

Usually sub-cool is the term used to describe the difference between the saturation temperature (boiling point) of water at the producer pressure and the actual temperature at the same place where the pressure is measured. The higher the liquid level above the producer the lower the temperature and higher is the sub-cool. However real life reservoirs are invariably heterogeneous therefore it becomes extremely difficult to achieve a uniform sub-cool along the entire horizontal length of a well.

As a consequence many operators, when faced with uneven stunted steam chamber development, allow small quantity of steam to enter into the producer to keep the bitumen in the entire wellbore hot hence keeping its viscosity low with the added benefit of transferring heat to colder parts of the reservoir along the wellbore. Another variation sometimes called Partial SAGD is used when operators deliberately circulate steam in the producer following a long shut-in period or as a startup procedure. Though a high value of sub-cool is desirable from a thermal efficiency standpoint as it generally includes reduction of steam injection rates but it also results in slightly reduced production due to a corresponding higher viscosity and lower mobility of bitumen caused by lower temperature.

Another drawback of very high sub-cool is the possibility of steam pressure eventually not being enough to sustain steam chamber development above the injector, sometimes resulting in collapsed steam chambers where condensed steam floods the injector and precludes further development of the chamber.

Operating the injection and production wells at approximately reservoir pressure eliminates the instability problems that plague all high-pressure steam processes and SAGD produces a smooth, even production that can be as high as 70% to 80% of oil in place in suitable reservoirs. The process is relatively insensitive to shale streaks and other vertical barriers to steam and fluid flow because, as the rock is heated, differential thermal expansion causes fractures in it, allowing steam and fluids to flow through.

This allows recovery rates of 60% to 70% of oil in place, even in formations with many thin shale barriers. Thermally, SAGD is twice as efficient as the older cyclic steam stimulation (CSS) process, and it results in far fewer wells being damaged by high pressure. Combined with the higher oil recovery rates achieved, this means that SAGD is much more economic than pressure-driven steam process where the reservoir is reasonably thick.

Invention and Development

The gravity drainage idea was originally conceived by Dr. Roger Butler, an engineer for Imperial Oil around 1969. But it wasn't until 1975 when Imperial Oil moved him from Sarnia, Ontario to Calgary, Alberta to head their heavy oil research effort that he pursued the concept. He tested the concept with Imperial Oil in 1980, in a pilot at Cold Lake, Alberta which featured one of the first horizontal wells in the industry, with vertical injectors. The latter were established to be inefficient by research at the Alberta Oil Sands Technology and Research Authority (AOSTRA) in the early '80s. This resulted in the first test of twin (horizontal) well SAGD, at their Underground Test Facility (UTF) in the Athabasca Oil Sands, which proved the feasibility of the concept, briefly achieving positive cash flow in 1992 at a production rate of about 2000 bbl/day from 3 well pairs.

The original UTF SAGD wells were drilled horizontally from a tunnel in the limestone underburden, accessed with vertical mine shafts. The concept coincided with development of directional drilling techniques that allowed companies to drill horizontal wells accurately, cheaply and efficiently, to the point that it became hard to justify drilling a conventional vertical well any more. With the low cost of drilling horizontal well pairs, and the very high recovery rates of the SAGD process (up to 60% of the oil in place), SAGD is economically attractive to oil companies.

Current Applications

This technology is now being exploited due to increased oil prices. While traditional drilling methods were prevalent up until the 1990s, high crude prices of the 21st Century are encouraging more unconventional methods (such as SAGD) to extract crude oil. The Canadian oil sands have many SAGD projects in progress, since this region is home of one of the largest deposits of bitumen in the world (Canada and Venezuela have the world's largest deposits).

The SAGD process allowed the Alberta Energy Resources Conservation Board (ERCB) to increase its proven oil reserves to 179 billion barrels, which raised Canada's oil reserves to the second highest in the world after Saudi Arabia and approximately quadrupled North American oil reserves. As of 2011, the oil sands reserves stand at around 169 billion barrels.

Disadvantages

As in all thermal recovery processes, cost of steam generation is a major part of the cost of oil production. Historically, natural gas has been used as a fuel for Canadian oil sands projects, due to the presence of large stranded gas reserves in the oil sands area. However, with the building of natural gas pipelines to outside markets in Canada and the United States, the price of gas has become an important consideration. The fact that natural gas production in Canada has peaked and is now declining is also a problem. Other sources of generating heat are under consideration, notably gasification of the heavy fractions of the produced bitumen to produce syngas, using the nearby (and massive) deposits of coal, or even building nuclear reactors to produce the heat.

A source of large amounts of fresh and brackish water and large water re-cycling facilities are required in order to create the steam for the SAGD process. Water is a popular topic for debate in regards to water use and management. As of 2005 Oil and Gas (including drilling) uses approximately 4.91% of total surface water and approximately 20.11% of ground water. Note that these numbers do not account for industrial uses such as injection, though the figures are also exclusive of recycled, reused and/or water returned to the water grid. Keep in mind that there has been drastic oil sands expansion since 2005 and one should consider this when estimated usage to date. The concern of using large amounts of water has little to do with proportion of water used, rather the quality of the water. Traditionally close to 70 million meters cubed of the water volume that was used in the SAGD process was fresh, surface, water. There has been a significant reduction in fresh water use as of 2010, where only around 18 million meters are used. Though the to offset the drastic reduction in fresh water use, industry has began to significantly increase the volume of saline, ground, water involved. Saline ground water, in Alberta, is obtained from wetlands and so by increasing the volume of water drained may leave cause for concern of marsh and wetland conditions as a result of the SAGD process. Relying upon gravity drainage, SAGD also requires comparatively thick and homogeneous reservoirs, and so is not suitable for all heavy-oil production areas.

Alternative Methods

Alternative enhanced oil recovery mechanisms include VAPEX (for Vapour Extraction), Electro-Thermal Dynamic Stripping Process (ET-DSP), and ISC (for In Situ Combustion). VAPEX uses solvents instead of steam to displace oil and reduce its viscosity. ET-DSP is a patented process that uses electricity to heat oil sands deposits to mobilize bitumen allowing production using simple vertical wells. ISC uses oxygen to generate heat that diminishes oil viscosity; alongside carbon dioxide generated by heavy crude oil displace oil toward production wells. One ISC approach is called THAI for Toe to Heel Air Injection.

Bibliography

Armistead, George: *Safety in Petroleum Refining and Related Industries*, New York: John G. Simmons & Co., 1950.

Bhattacharya, K. G. and A. K. Talukdar: *Catalysis in Petroleum and Petrochemical Industries*, Alpha Science International, Delhi, 2005.

Black, Brian: *Petrolia: The Landscape of America's First Oil Boom*. Baltimore: Johns Hopkins University Press, 2000.

Chester, Edward W.: *United States Oil Policy and Diplomacy: A Twentieth-Century Overview*. West-port, Conn.: Greenwood Press, 1983.

Chisholm, Archibald H. T.: *The First Kuwait Oil Concession Agreement: A Record of the Negotiations,* 1911-1934. London: Frank Cass, 1975.

Davenport, E. H. and Sidney Russell Cooke: *The Oil Trusts and Anglo-American Relations*. Macmillan, 1924

Diwan, Parag and Ashutosh Karnatak: *Handbook of Natural Gas Technology and Business*, Pentagon Energy Press, Delhi, 2009.

Eaton, Rev. S.J.M.: *Petroleum, a History of the Oil Region*, J.P. Skelly & Co., Philadelphia, 1866.

Engler, Robert: *The Politics of Oil: A Study of Private Power and Democratic Directions*. New York, Macmillan, 1961.

Fischer, Louis: *Oil Imperialism: The International Struggle for Petroleum*. New York: International Publishers, 1926.

Garverick, Linda: *Corrosion in the Petrochemical Industry*. Materials Park, Ohio: ASM International, 2000.

Giddens, Paul H.: *The Birth of the Oil Industry*, The MacMillian Company, New York, 1938.

Hartshorn, J. E.: *Oil Companies and Governments: An Account of the International Oil Industry in Its Political Environment*. London: Faber and Faber, 1962.

Hyne, Norman J.: *Nontechnical Guide to Petroleum Geology, Exploration, Drilling, and Production*. PennWell Corporation, Delhi, 2001.

Jentleson, Bruce: *Pipeline Politics: The Complex Political Economy of East-West Energy Trade.* lthaca: Cornell University Press, 1986.

Klare, Michael: *Blood and Oil: The Dangers and Consequences of America's Growing Petroleum Diplomacy.* Metropolitan Books/ Henry Holt, 2004.

Klare, Michael: *Resource Wars: The New Landscape of Global Conflict.* Metropolitan Books / Henry Holt, 2001.

Lubell, Harold: *Middle East Oil Crises and Western Europe's Energy Supplies.* Baltimore: Johns Hopkins University Press, 1963.

Marsh, Steve: *Anglo-American Relations and Cold War Oil: Crisis in Iran.* New York: Palgrave MacMillan, 2003.

McNaugher, Thomas: *Arms and Oil: US Military Strategy and the Persian Gulf.* Brookings Institution, Londong, 1985.

Naderpour, N and F Hematti Kordmahaleh: *Petroleum Geosciences*, SBS Pub, Delhi, 2009.

Odell, Peter R.: *Oil and World Power: Background of the Oil Crisis,* New York: Viking Penguin, 1986.

Olien, Diana Davids: *Oil in Texas: The Gusher Age, 1895–1945.* Austin: University of Texas Press, 2002.

Penrose, Edith, and E. F. Penrose: *Iraq: International Relations and National Development.* London: Ernest Bern, 1978.

Randall, Stephen J. *United States Foreign Oil Policy, 1919–1948: For Profits and Security.* Kingston, Ontario, 1985.

Rister, Carl Coke: *Oil! Titan of the Southwest.* Norman: University of Oklahoma Press, 1949.

Sampson, Anthony: *The Seven Sisters: The Great Oil Companies and the World They Shaped.* New York, 1975.

Simmons, Matthew R.: *Twlight in the Desert: The Coming Saudi Oil Shock and the World Economy.* 2005

Tribedi, S. : *Proceedings of National Convention of Environment in India, Challenge for the 21st Century*, India, 1994.

Venn, Fiona: *Oil Diplomacy in the Twentieth Century.* New York, 1986. Pioneering study stressing oil's strategic importance.

Vervalin, Charles H.: *Fire Protection Manual for Hydrocarbon Processing Plants.* Houston: Gulf Publishing Co, 1985.

Watkins, Eric: *The Middle Eastern Environment*, London, The British Society for Middle Eastern Studie, 1995.

Wendel, Charles H.: *The Allis-Chalmers Story.* Sarasota, Florida: Crestline Publishing Co., 1988.

Yergin, Daniel: *The Prize: The Epic Quest for Oil, Money, and Power.* New York: Simon and Schuster, 1991.

Index

❑❑❑